GET IN THE GAME

GET IN THE GAME

AN INTERACTIVE INTRODUCTION TO SPORTS ANALYTICS

TIM CHARTIER

Illustrated by Ansley Earle

The University of Chicago Press   Chicago and London

The University of Chicago Press, Chicago 60637
The University of Chicago Press, Ltd., London
© 2022 by Timothy P. Chartier
Illustrations by Ansley Earle. © 2022 by The University of Chicago
Published 2022
Printed in the United States of America

31 30 29 28 27 26 25 24 23 22   1 2 3 4 5

ISBN-13: 978-0-226-81114-7 (paper)
ISBN-13: 978-0-226-81128-4 (e-book)
DOI: https://doi.org/10.7208/chicago/97802268112842.001.0001

Library of Congress Cataloging-in-Publication Data

Names: Chartier, Timothy P., author. | Earle, Ansley, illustrator.
Title: Get in the game : an interactive introduction to sports
analytics / Tim Chartier ; illustrated by Ansley Earle.
Description: Chicago : University of Chicago Press,
2022. | Includes bibliographical references.
Identifiers: LCCN 2022004663 | ISBN 9780226811147
(paperback) | ISBN 9780226811284 (ebook)
Subjects: LCSH: Sports—Statistics.
Classification: LCC GV741 .C455 2022 |
DDC 796.021—dc23/eng/20220201
LC record available at https://lccn.loc.gov/2022004663

♾ This paper meets the requirements of
ANSI/NISO Z39.48-1992 (Permanence of Paper).

CONTENTS

1

UNFORGETTABLY UNBELIEVABLE

You've seen it, but not every day. You're watching a sporting event and the seemingly impossible happens—a diving catch, a half-court shot, a lunge to the finish line. And the crowd erupts. It's as if we are cheering humanity's ability to overcome—in the last moment, with the last attempt, against incredible odds.

The memorable event may be a topic of discussion with friends or on social media for a day or a season. But most such highlights fade from memory before long, overshadowed by the next memorable accomplishment. Only occasionally does the unbelievable become the unforgettable. When something remarkable occurs, can we distinguish between the momentarily memorable and the truly

historic? The historic is not likely to happen again for a long time, if ever.

The unforgettable is often improbable. So understanding the probability of a surprising event is a clue to its historic significance. How can we compute those odds? And can we do so quickly? Is it possible to see a remarkable accomplishment and by the end of a commercial break know that we likely saw an unforgettably improbable moment?

In this book, we'll explore how to estimate the probability of events as a way to get in the game with sports analytics. We'll focus on what counts as historic, but you can apply the tools we'll use to more everyday questions: Who might win tonight's game? Can last week's star player continue at their current level of play? Who might be ready to slay a giant?

That could send you down a rabbit hole. A lot of math, statistics, and computer science topics are relevant to such analysis. One could binge online videos, read hundreds of articles, or learn the nuances of complex and precise analytics tools. But looser estimates can take us a long way toward evaluating the improbable. Here we'll find insights in flipping a coin or rolling a die. We'll learn what are often called "back of the envelope" computations, quick calculations to get a sense of a quantity, if not necessarily an exact value.

So grab a pencil and paper and your phone or a computer—you'll want the calculator for some quick arithmetic, and you'll sometimes need to download data. That, plus curiosity, is all you'll need.

Let's start by looking back a hundred years to the 1920 Major League Baseball season. Babe Ruth was synonymous with home runs. Grabbing stats from baseball-reference.com, we see that in 1920 Ruth hit 54 home runs. The league average that year for home runs by a team was 39. In other words, Ruth hit 54/39, or about 1.4 times more homers than the average MLB *team*. Now jump ahead to 2019, when the league average for home runs by a team was 226. Repeating Ruth's feat, hitting 1.4 times the team average, would have meant hitting 313 home runs in the 2019 season—almost two home runs per game, 4.5 times the best rate ever.

Suppose you notice that Babe Ruth was a leftie. So you check the 2020 Yankees roster and find that about 21% of those players threw

leftie. That seems higher than the general population. But it's just one statistic for one team in one season. To get a larger sample size, you could download data from SeanLahman.com, which has stats for more than twenty thousand MLB players, including whether they threw left- or right-handed.

A quick spreadsheet calculation reveals that between 1871 and 2019 about 20% of players threw left-handed, which is close to our 2020 percentage for the Yankees. Are a fifth of your friends left-handed? It's unlikely. According to the *Washington Post*, about 10% of the people in the world are lefties.

So far we've used a calculator to compare Babe Ruth's 1920 home run production with that of the average team, and a spreadsheet to tally the MLB's historical percentage of lefties, which we then compared with that of the general population. Now we'll look at another enduring name in sports, NFL wide receiver Jerry Rice, and gain insight from graphing.

Graphs often make outliers obvious. When we plot the number of career receptions for the top ten players in this category on a horizontal axis, look how much longer the horizontal bar is for Jerry Rice than for the others. This difference visually underscores Rice's status as an outlier even among Hall of Fame wide receivers.

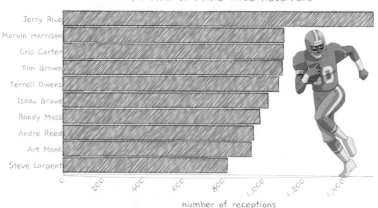

NFL Hall of Fame Wide Receivers

number of receptions

That's a remarkably unforgettable career stat. There are also re-markably unforgettable moments in single games. In the 2015 FIFA Women's World Cup final, the USA beat Japan 5–2, and Carli Lloyd scored a hat trick (three goals) in the first sixteen minutes. Unbe-lievably fast? How much time did it take to complete hat tricks in other World Cup games? In this next graph, the shorter the bar, the faster a hat trick was scored. The historic nature of Lloyd's hat trick is again visually obvious: her bar is roughly half the length of the second shortest bar (which is very close in length to the third and fourth shortest).

Without much math, then, we've gained insight on what makes an athletic performance historic. A critical element is the data. Babe Ruth hit 1.4 times more homers than the average team in 1920. Jerry Rice was similarly far ahead of his nearest career competitor. And Carli Lloyd scored her hat trick in half the time of the next fastest effort. If you dig into the data, you'll see that the other World Cup hat tricks not only took more time but did not occur in a final. In each case, the numbers, calculated or visualized, emphasize the un-believable improbability of a season, career, or game.

Playbook

An **outlier** is a data point that differs significantly from other comparable data. We can visually spot Babe Ruth's 1920 home run total as an outlier by graphing the number of homers hit by the top five home run hitters that year.

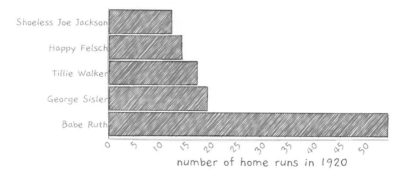

Note that this definition leaves the analyst to decide if a difference is significant.

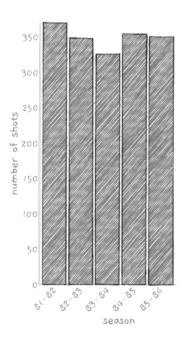

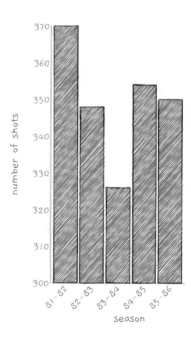

Graphs can help us spot outliers, but beware: they can also be misleading. These two graphs depict the same data—the number of shots by Wayne Gretzky in five of his nine MVP seasons. So why do they look different? The vertical axis of the graph on the left begins at 0 (like the horizontal axis of the 1920 home run graph), whereas the vertical axis of the graph on the right begins at 300. In the second graph, the difference in shots between 1981–82 (370) and 1983–84 (326) looks much bigger than it actually is.

A response to all this analysis can be "I just want to watch the game." But watching a game can still allow a lot of time for quick analysis. Take the NFL. An average game lasts well over three hours. How much actual action occurs? In football, the clock often keeps running as teams huddle  or take formation. According to statistics site FiveThirtyEight, the average amount of action in an NFL game is eighteen minutes. And how much of the three hours is commercials? Fifty minutes! With such limited opportunity to jump up and root for the home team, we might feel we're putting down roots, becoming true couch potatoes. With the tools of this book, when someone says, "Wow, that's amazing!" you'll be able to flip a coin, roll a die, jot down a few numbers, or plot some data and reply, "Want to know just *how* amazing?"

Highlight reels could be renamed outlier reels. Within the stream of daily and weekly highlights are moments that will transcend time. Analytics can help us recognize the historic. We've begun assembling our analytics toolbox. In the coming chapters, we'll add to it so we can more fully appreciate the unforgettably unbelievable and get in the game of sports analytics.

Personal Training—Workout #1

Each chapter will introduce new analytic tools and end with a workout where you can try them out. So, get warmed up—here is your first personal training workout!

Any world record in sports is an outlier performance. But some world records stand significantly longer than others. Let's turn to swimming and look at the progression of records in the women's 1500 meter freestyle. Look online for the data and you'll find that since 1922 new records have been set forty-five times. For this challenge, collect the dates these records were set, then determine how many days each record remained unbeaten. You'll note that the longest-standing record (15 minutes, 52.10 seconds) was set by Janet Evans of the United States on March 26, 1988, and held until June 17, 2007.

Question: How does the longevity of Evans's 1988 record compare to that of other world records in this event? What is the best way to represent this data to reveal outliers?

After you've completed your work on this personal workout, turn to the back of the book for a discussion of the question.

2

SHOOT 3'S LIKE STEPH CURRY

On February 27, 2013, highlight reels played and replayed clips from the previous night's game at Madison Square Garden. The focus was scoring by the losing team—and mainly by one player. Why? Stephen Curry of the Golden State Warriors had racked up 54 points against the Knicks. The first quarter had not been promising, as Steph scored only 4 points, making two out of five two-point attempts and missing his only three-point shot. But the unforgettable can quickly follow the unimpressive. Curry's second-quarter total was 23 points: three of five two-point shots, four of five three-point shots, and all five free throws. Impressive? Indeed! Improbable? How might we decide?

Part of what makes an event unforgettable involves comparisons across history, as we saw in the last chapter with Carli Lloyd and Jerry Rice. Only seven players in NBA history had scored more points in one game at Madison Square Garden. Who? It's like an NBA Who's Who—Wilt Chamberlain, Kobe Bryant, Michael Jordan, Rick Barry, Bernard King, Elgin Baylor, and Richie Guerin. Curry accomplished several remarkable feats that night—54 points in a game, 23 points in a quarter, and 11 of 13 three-pointers.

Consider that last statistic: Curry made 11 of 13, or 84.6% of his three-point shots. That's a very high percentage—super high. Only twelve NBA Hall of Famers have a higher career shooting percentage from the free throw line, and no one has sustained such accuracy from the three-point arc.

Curry is among the players who have transformed distance shooting in the NBA. To see this, let's compare Steph to his Warriors coach, Steve

Kerr. In his five seasons with the Chicago Bulls in the 1990s, Kerr, also a distance shooter, attempted 898 three-pointers, making 430. In the 2012–13 season, Curry made 272 of 600 three-point shots. These convert to similar percent-ages: Kerr's 430 of 898 is about 47%, and Curry's 272 of 600 is about 45%.

How can we analyze the improbability of Curry's 11 of 13 shoot-ing? Steph, of course, played the game only once. You'll play it mul-tiple times. In a sense, you get to have a shooting contest with the back-to-back NBA MVP. No need to find a court and a basketball. All you need is a penny! You'll compete, and analyze Curry's perfor-mance, by flipping over three-pointers.

Playbook

A **probability**—the likelihood that an event will occur—is repre-sented by a number between 0 and 1. An event with a probability of 0 definitely won't happen; an event with a probability of 1 definitely will happen. An event may consist of a single outcome (like rolling a particular number with one die) or a group of outcomes (like rolling a number with two dice, or hitting some percentage of attempted three-pointers).

We will often use the number of ways dice can be rolled to model more complex situations. If you roll one six-sided die there are six possible outcomes, each with an equal, one-in-six probability (1/6 = 0.167, or 16.7%). If you roll two dice, there are six possible out-comes for each die—for example, you might roll a 4 with the first die and a 3 with the second die (which we will write as 4-3). That means there are thirty-six possible combined outcomes (6×6, or 6^2).

Each of these outcomes is equally likely. So then, what is the probability of a particular event, for example, rolling a 3? Out of thirty-six total ways to roll two dice, just two yield a sum of 3: 1-2 and 2-1. The probability of rolling a 3 with two dice is the number of outcomes in which the event occurs (2) divided by the total number of possible outcomes (36): 2/36 = 0.055, or 5.5%.

To determine the improbability of Curry's three-point fest, we'll use his 2012–13 three-point shooting percentage of 45.3% (which you can find with a quick Google search). Let's assume that a flipped coin is equally likely to land heads or tails. So the chance of flipping heads is 50%. You'll flip to "shoot" and score by getting heads—meaning your accuracy is slightly greater than Curry's season average (50% versus 45.3%). But for your one-on-one shooting contest, Steph's turn is fixed as his far better than average February 27 performance. You flip your coin thirteen times. If you flip heads eleven times, you and Steph tie. If you flip more than eleven heads, you win. Else, Steph wins.

Worksheet: One-on-One with Steph Curry

We're ready to play our game. Curry's results are locked in; he's made 11 of 13 three-point attempts. Now you need to take your 13 shoots.

1 Flip a coin.
2 If you get heads, you make the shot; tails, you miss.
3 Repeat the coin flip 13 times and keep track of your progress with the chart below.

Did you beat Steph by making more than eleven shots? If so, you're lucky. Play again (and again), and you'll likely see that your win isn't easily repeated. In fact, you have less than a 1% chance of getting more than eleven heads in thirteen flips.

Of course, Curry is such a great shooter. Shouldn't he win every shooting contest? On a physical court, almost certainly, but you are flipping a coin. Your shooting percentage is 50%, higher than his. Yet you'll rarely beat him.

How do I know you are less than 1% likely to win this shooting contest? To establish that number, I also flipped a coin—just many, many more times.

Grab a dollar and cash it in for pennies. Give the pennies to a hundred people, and ask each person to flip their coin thirteen times. How many people get more than eleven heads? Given that the underlying probability is less than 1%, maybe no one.

More games will mean a better estimate. Go big and ask a hundred thousand people to each flip a coin thirteen times. You'd find about 170 people get more than eleven heads and around 1,000 will get exactly eleven heads (meanwhile, about 42,000 will get 6 or 7 heads). So your estimated probability of beating Steph's shooting would be 170/100,000 = 0.17%, and the probability of matching Steph's performance would be 1,000/100,000 or 1%.

We now have an important tool for our analytics toolkit. This process, called **simulation**, can uncover the improbable. Simulation helps analyze events. For us, the event was hitting eleven of thirteen shots. We made some assumptions. We were shooting 50% with our coin flipping, which is slightly high. We also assumed that every shot is the same. But is it? Steph was playing against other players, who would have made varying decisions as the game unfolded. Also, he missed his only three-point attempt in the first quarter, then hit 4 of 5 in the second. Do any of these conditions influence his second-half shooting? Our model assumes they do not.

Playbook

Our model for Steph Curry's shooting assumes the same probability for every shot, meaning the outcome of one shot doesn't influence the outcome of another shot. When the occurrence of one event does not affect the probability of another, we say the outcomes are **independent**. This is true of coin flips and dice rolls, but if Steph makes one shot, is he more or less likely to make the next? In this model, we assume it doesn't make a difference. For a variety of reasons this isn't precisely true. Offenses and defenses change in response to made shots, and it takes more complex models to integrate such subtleties. Still, we can make useful estimates assuming the independence of events, as a variety of models throughout this book will do.

So what can our new tool tell us? Why aren't we winning our shooting contests? Curry's massive advantage in this one-on-one contest is our fixing his shots to one improbable performance. Let's assume that, as you watched Curry warm up in Madison Square Garden on that February night, you somehow knew he'd shoot thirteen three-pointers. You'd have every right not to believe he would hit eleven of them. It's improbable, and that unlikeliness is what made that night so memorable.

Many of the best NBA and WNBA guards have shooting percentages over 40%. Having a coin handy can help you get a sense of the improbability of their shooting sprees.

Take another Steph Curry game. He made 13 of 17 three-pointers against New Orleans in November 2016 (that season he hit 41%, close enough to his 2012–13 percentage that we can use the same model). Was this performance likely? Grab your coin and flip seventeen times. Whether or not you flip thirteen heads, try it again and again. In chapter 4, we'll learn a formula that demonstrates that getting 13 heads in 17 flips is almost twice as likely as getting 11 heads in 13 flips. But two times a 1% probability is still not very likely. This quick flipping experiment signals that the 2016 shooting spree is also likely more than just a highlight.

But what about other situations in sports, where the probability isn't close to 50%? We'll learn about this and much more in coming chapters. For now, find a good spot for that penny so it's handy for your sports watching.

Analytic Toolkit

What if you don't have a coin handy? Just type "flip penny" into Google and you'll have an online, virtual coin flipper.

(And what if you don't know 100,000 people you can ask to flip coins? Either step up the social networking or wait for the helpful formula in chapter 4.)

Personal Training—Workout #2

Stephen Curry won the 2020–21 NBA scoring title, averaging 32.0 points per game. But not every game was a festival of swishes. In early January, Curry made only 2 of 16 shots against the Raptors, his worst shooting night to date.

Question: How improbable was it for Curry to go 2 for 16 from the field in a game that season?

Approach 1: Look up Stephen Curry's 2020–21 regular season statistics. Can you use a coin to estimate his shooting?

Approach 2: How can you estimate Curry's two-point shooting percentages with multiple coin flips? How about his three-point shooting?

Try working out an answer, then check out the discussion in the back of the book.

3

DICEY HITTING STREAK

Sibling rivalries are nothing new. Cleopatra and Ptolemy XIII clashed over their father's wish that they jointly rule in Egypt; in the end, Ptolemy drowned in the Nile while fighting to regain his throne. In the 1920s, Adolf and Rudolf Dassler created a shoe company in their mother's laundry room. Success led to tension and their splitting in the aftermath of World War II. Adolf started Adidas, while Rudolf created Puma.

There are also the sibling rivalries in sports. In tennis, Serena and Venus Williams each hold an impressive number of Grand Slam titles. In football, NFL quarterbacks Eli and Peyton Manning have both won Super Bowls. In basketball, Reggie Miller is known

for scoring eight points in nine seconds in the 1995 NBA playoffs, while Cheryl Miller was a four-time All-American and one of the top rebounders in NCAA history. In boxing, Michael Spinks won all but one of his thirty-two fights; brother Leon was less successful but could say one thing Michael couldn't: "I beat Muhammad Ali!"

A sibling rivalry for the record books is that of Joe and Dom DiMaggio. Their two oldest brothers joined their father Giuseppe as fishermen. Joe, Dom, and older brother Vince became major league baseball players via the sandlots of San Francisco. Vince DiMaggio was a two-time All-Star, but we'll focus on Joe and younger brother Dom and their still unforgettable hitting streaks.

First, let's check the stats. Joe DiMaggio had a career batting average of 0.325, with 361 home runs and 1,537 RBIs. Over thirteen seasons, he had only 369 strikeouts. His Hall of Fame career included thirteen All-Star appearances, nine World Series titles, and three American League MVPs. Dom hit 0.298 over his career, with 87 home runs and 618 RBIs. In eleven seasons, he had 571 strikeouts and was a seven-time All-Star.

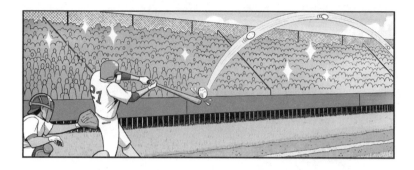

Our duel takes place in the 1940s. In 1941 Joe DiMaggio posted a 56-game hitting streak. In 1949, Dom had a 34-game hitting streak. Both are impressive: Joe's streak is still the Major League Baseball record eighty years later, and Dom's remains the longest in Boston Red Sox history.

In chapter 2, we played Steph Curry one-on-one. Because his three-point shooting percentage for the 2012–13 season was close to 50%, we could get a sense of whether a memorable performance

was an outlier by flipping a coin. No one hits close to 50% in Major League Baseball. Joe hit 0.357 in 1941. Dom hit 0.307 in 1949. So, how can we play baseball? This is where things get dicey. A simple roll of a die will let us hit like a DiMaggio! When we roll a six-sided die, there are six possible outcomes, all equally likely. So rolling a 1 or a 2 represents a third of the possibilities. In other words, we have a 33% chance (0.333) of rolling a 1 or 2, which is between Joe's and Dom's batting averages. While we could roll a die for every at-bat, let's expand our analytics toolkit to compute the DiMaggios' probabilities.

....

Playbook

To analyze the improbability of the DiMaggio brothers' hitting streaks, we need two more tools. First, if the probability of an event is p then the probability of that event not happening is $1 - p$. Remember, a probability is represented by a number between 0 and 1. An event with a probability of 1 will certainly happen, whereas an event that can't happen has a probability of $1 - 1 = 0$. In between those extremes, the probability of, for example, flipping two tails in two coin flips is 0.25 (1 of 4 possibilities: T-T). The probability of not flipping two tails in two flips, then, is $1 - 0.25 = 0.75$ (3 of 4 possibilities: H-H, H-T, T-H).

Second, to calculate the probability of one event happening and then another event happening, we can multiply their probabilities, as long as the events are independent. So, the probability of flipping two tails in a row and then, on the next two flips, not flipping two tails in a row is $0.25 \times 0.75 = 0.1875$.

....

Let's assume a player has a 0.250 batting average. (Note: we'll keep the third decimal digit even if it's a zero, since a 0.250 batting average is often referred to as "hitting 250.") There are multiple ways to get at least one hit in a game but only one way to go hitless, which is helpful. If the probability of a hit is 0.250 for each at-bat, then the probability of not getting a hit is $1 - 0.250 = 0.750$. So the probability of not getting a hit in two straight at-bats is 0.750^2, and three straight at-bats is 0.750^3. If our player comes to the plate four times in each game, then the probability of going hitless in a game is $0.750^4 = 0.316$. Since getting at least one hit is the opposite of going hitless, the probability of getting at least one hit is $1 - 0.316 = 0.684$. So the probability of getting a hit in two straight games is 0.684^2; in three straight games, 0.684^3; and so on.

We can use the same analysis with Joe and Dom DiMaggio's batting averages. We'll assume each brother came to the plate four times in each of his games. Joe's 1941 batting average was 0.357, so the probability of his not getting a hit was $1 - 0.357 = 0.643$ each time he walked to the plate. And the probability he would be hitless in four consecutive at-bats would be $0.643^4 = 0.171$.

The probability of Joe DiMaggio getting at least one hit in a game, then, would be $1 - 0.171 = 0.829$. Since $5/6 = 0.833$, you can simulate the odds with a single die: roll 1 through 5 and you get a hit; roll a 6 and you don't. Can you hit like Joe and roll fifty-six straight times without rolling a 6?

Joltin' Joe DiMaggio

Worksheet: Joe DiMaggio Hitting Contest

Joe DiMaggio got a hit in 56 straight games.

1 Roll a die.
2 If you get 6, you end the game hitless and Joe wins the contest. Else, you get a hit in one game.
3 Keep track of your progress with the chart below, through 56 games or until your streak ends.

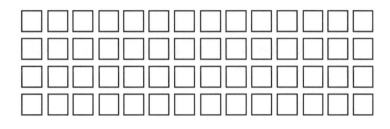

It's unlikely you beat Joe in the hitting contest. If a thousand people try, it's likely none of them will beat him either. How unlikely? The probability of *not* rolling a 6 on one roll (that is, the probability of rolling a 1, 2, 3, 4, or 5) is 5/6 (0.833). The probability of not rolling a 6 on two rolls is $0.833 \times 0.833 = 0.833^2$. The probability of not rolling a 6 on three rolls is 0.833^3. And the probability of not rolling a 6 on fifty-six straight rolls is 0.833^{56}, or about 0.000036. Calculating the inverse of that number—1/0.000036—gives us the odds for matching Joe's 1941 streak: about 1 in 28,000. Let's compare that to the odds of hitting in thirty-four straight games (the length of Dom's 1949 streak), which would be 0.833^{34}, or about 1 in 500.

Playbook

The assumption of four at-bats per game is a simplification. For one thing, walks do not count as at-bats. Ted Williams, who was walked much more frequently than Joe, set a record in 1949 of 84 straight games in which he got on base via a hit or a walk.

Our computation estimates the probability of Joe's hitting streak beginning at a given moment. We'll see in chapter 10 that consistency can also be integrated into such models.

....

So far, we've used the same batting average for Joe's and Dom's hitting streaks. To better compare the improbability of the brothers' accomplishments, we need to use their individual batting averages. Let's begin with Joltin' Joe. We've already computed the probability of Joe DiMaggio getting at least one hit in a game with four at-bats as 0.829. So, the probability of his getting at least one hit in two consecutive games would be 0.829^2; in three games, 0.829^3; and in 56 games, 0.829^{56}. Calculating $1/0.829^{56}$, we see that this probability is about 1 in 36,000.

Dom's batting average in 1949 was 0.307. So his probability of going hitless in a game would be $(1 - 0.307)^4 = 0.231$, making the probability of getting at least one hit $1 - 0.231 = 0.769$. The proba-

bility of Dom hitting in thirty-four straight games, then, is 0.769^{34}, or about 1 in 7,500.

So far, Dom's hitting streak looks less unlikely than Joe's—but of course, it was shorter. What if he'd kept it up? The formula is easy enough to adapt. Dom's chances of matching his brother's 56-game hitting streak would be 0.769^{56}, or about one in 2.5 million. Comparing to Joe's 1 in 36,000 odds, we find that it was over 65 times more likely for Joe than Dom to hit in fifty-six straight games.

Our analysis helps explain why Joe DiMaggio's record has stood for eighty years. Joe's 1941 batting average (0.357) was higher than his career average, 0.325;

using that number in our analysis would reduce his chance of a 56-game hitting streak to about 1 in 460,000. Out of approximately twenty thousand MLB players since 1871, Joe's is among the top fifty career batting averages, and his 1941 number is among the top 210 single-season batting averages. So if you were in the stands, watching a black-and-white TV, or listening to the radio during Joe's hitting streak, our analysis confirms you would have been right to think you were witnessing a feat that was not only unforgettably improbable but quite possibly unrepeatable.

If you see an MLB hitting streak, use our formulas or grab a die and see if you can repeat the player's success. So next to that penny from the last chapter, you may want to put a die. If you prefer more exact probabilities, add some scrap paper to run through the calculations.

Analytic Toolkit

What if you don't have a die handy? Simply type "roll dice" into Google and you'll have an online, virtual dice roller. The default, as shown here, is one six-sided die. Click "Roll" again and again and see if you can have a hitting streak like Joe DiMaggio.

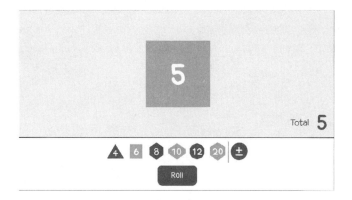

You can also use Google's calculator. Type "(5/6)^56" into Google's search bar; Google will do the math and return the value of 0.0000368002.

Personal Training—Workout #3

During the 2007–8 season, the Houston Rockets had a 22-game winning streak, the second-longest in NBA history at the time. When the streak began, in late January, the team had won just over half its games and was tenth in the Western Conference. Halfway through the streak, the Rockets lost their leading scorer and only All-Star, Yao Ming. Still, they kept on winning, and when the streak ended, they led the conference.

Given their record when the streak began, let's assume the Rockets had a 50% chance of winning any game.

Question: Assuming a 50% chance of winning any game, how improbable was it for the Houston Rockets to win 22 straight games?

Work out your answer, then check out the discussion in the back of the book.

RACKING UP THE WINS

Winning streaks can be unforgettable. The United States kept the America's Cup in sailing for 132 years. In 1967 Richard Petty won ten NASCAR races in a row. Boxer Rocky Marciano entered the professional ring in 1947 and retired in 1956 with an undefeated 49–0 record. A more recent streak began in the fall of 2014 when the University of Connecticut women's basketball team, already two-time defending NCAA champions, began winning even more—a lot more! For more than two years, they didn't lose. Across three seasons, they won 111 straight games, including two more national championships. In the course of the streak, they won by ten or more points in all but three games.

The Huskies, now four-time defending champions, entered the 2017 March Madness tournament with over a hundred straight wins. Las Vegas rated them 21.5-point favorites, with a 97.9% chance of victory, in their Final Four semifinal game against Mississippi State. A $200 bet on Mississippi State would return $4,000 if the Bulldogs won.

You might not have bothered tuning in until you heard the game was in overtime, which was a rollercoaster. With twelve seconds to go the game was again tied. Mississippi State held the ball, and a shot at the buzzer by Morgan William sailed through the air, seemingly in slow motion, sinking into the net. Game over. Streak over. History made!

Such an amazing streak is clearly improbable. But can we say just how improbable? If you were watching the University of Connecticut women notch their 50th, 100th, or 111th win, how could you estimate the odds?

In the last chapter, we analyzed Joe DiMaggio's hitting streak. To attempt a similar analysis here, we need to specify the probability of the Huskies winning each of their games. Now, we could vary the probabilities from game to game. If you are 50% likely to win a given game and 30% likely to win the next game, then the likelihood of winning both games would be $0.5 \times 0.3 = 0.15$, or 15%. But determining distinct probabilities for

111 games would certainly be tedious and possibly quite difficult. The Huskies won all but three of those 111 games by ten or more points, so it seems safe to say the probability of each win would be high. To get a broad sense of the streak's improbability, let's assume the same high probability of a Husky win for every game.

Recall that odds-makers gave the Huskies a 97.9% chance of beating Mississippi State. Let's start off a tad more pessimistic and assume a consistent 95% chance of winning each game. The Huskies' likelihood of winning 111 straight games would then be 0.95^{111}, about 0.34%. That seems low, given their many convincing wins. So let's try the Vegas odds: assuming a 97.9% chance of winning each game, the likelihood of a 111-game winning streak becomes 0.979^{111}, or just under 9.5%. Let's roll dice and allow only one in 6^3 options to signal a loss, giving you a 99.5% chance of winning a game. With such odds of winning, how long will your winning streak be?

....

Worksheet: UConn Winning Streak Contest

The UConn women's basketball team won 111 straight games. How many can you win?

1 Roll three dice (or one die three times) to see if you win or lose a game.

2 If you get three ones, you lose the game; else you win.

3 Repeat this until you lose a game or win 111 games. Tally your winning streak with the chart below.

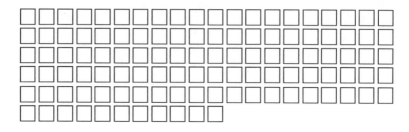

Did you achieve a long winning streak? Probably. With three six-sided dice, there are $6^3 = 216$ possible outcomes. Since only one of those outcomes counts as a loss, the probability of winning each game is 215/216, or about 99.5%, and your probability of winning 111 straight games is $(215/216)^{111}$, or just under 60%. Could UConn have enjoyed such a high probability? Possibly. Our 97.9% probability was based on the bookmakers' odds for the semifinal game against Mississippi State. The teams taking part in March Madness are stronger than average, so the Huskies' likelihood of winning many of their games those three seasons was easily higher.

With so many convincing wins, the Huskies could almost have seemed destined to rack up wins. Our computations demonstrate the importance of a team's dominance in a long winning streak. A team that is 90% likely to win each game has 1 chance in about 120,000 of winning 111 straight games. For such a winning streak to have at least a 50–50 likelihood, the probability of winning has to be above 99%. The Huskies' 111-game winning streak is unforgettable, and so was their dominance.

We have now explored two streaks. For the Huskies, it was winning games. For Joe and Dom DiMaggio, it was hitting in consecutive games. Let's analyze one more streak—Elena Delle Donne's free throw shooting. Delle Donne is the WNBA's all-time leader in free throw percentage, at 0.9390. From 2013 to 2021, she shot 1,018 free throws and missed only 62; in 2019, she missed only 3 in 117 attempts. In 2017 she tied the WNBA record for longest single-season streak, sinking 59 consecutive free throws. What is the probability of shooting 59 straight free throws? Assuming Delle Donne's career average of 93.84% holds for each shot, 0.9384^{59}, or about 2.3%. If you assume only 90% free throw shooting, the feat becomes ten times less likely.

....

Playbook

Using formulas is usually faster than counting possibilities by hand. The US women's gymnastics team competes with four gymnasts in the team competition. Assuming no ties, in how many ways can the four gymnasts finish first, second, third, and fourth on the balance beam? We have four choices for first, which leaves three choices for second, which leaves two choices for third, which leaves only one choice for fourth: $4 \times 3 \times 2 \times 1 = 24$ different ways. This can be extended. Suppose we have eight swimmers in the Olympic finals for the women's 50 meter freestyle. How many ways can the swimmers come in first, second, third, and fourth? This time we have eight choices for first, seven for second, six for third, and five for fourth:

8 × 7 × 6 × 5 = 1,680. In these cases, we are placing items (gymnasts or swimmers) in an order. When we select from a set of options and order matters, we are dealing with **permutations**. Selecting four items from eight options and placing them in order is denoted P(8,4).

When we select from a set of options and order does not matter, we are dealing with **combinations**. For example, the number of ways nine finalists in the 100 meter dash can finish first, second, and third is P(9,3), or 9 × 8 × 7 = 504. But in the Olympic trials, each of those top three finishers makes the team. If we want to know how many teams of three can be selected, the order in which they finish doesn't matter. A team made up of player 1, player 2, and player 3 is the same as a team made up of player 2, player 1, and player 3. So of those 504 permutations, we need to remove those that are just re-orderings of the same three items. How many ways can three items be ordered? P(3,3) = 3 × 2 × 1 = 6 ways. So the number of possible teams is P(9,3)/P(3,3) = 504/6 = 84, denoted C(9,3).

The formulas for permutations and combinations can be written more simply using the **factorial** (a number followed by an exclamation point): $n! = n \times (n - 1) \times (n - 2) \ldots \times 2 \times 1$ (so 3 × 2 × 1 in the preceding example can be denoted 3!). Here's the general formula for permutations, the number of ways you can choose r items from a group of n, when order in which they're chosen matters:

$$P(n,r) = n!/(n - r)!$$

And here's the general formula for combinations, C(n,r) (said "n choose r"), where the order in which the r items are chosen does not matter:

$$C(n,r) = n!/((n - r)! \times r!)$$

We never found an exact probability for Steph Curry's shooting in chapter 2. There is only one way to win 111 straight games but multiple ways to make eleven out of thirteen shots. You can miss two, then make eleven, or make eleven, then miss two, or mix up hits and misses. How many ways can one do this? Armed with $C(n,r)$, we can extend our analytics toolkit and tackle this question.

To use this tool, we need two things. First, there can only be two outcomes. In this case, Curry could make his shot or miss his shot. Second, we need to specify the probability of each outcome. Steph's three-point shooting average for the 2012–13 season was 45.3%. So the probability of making a shot is 0.453. Since there are only two options, the probability of missing a shot is $1 - 0.453$.

When you have two outcomes and know the probabilities of each, you have a Bernoulli trial, named after Swiss mathematician Jacob Bernoulli. Applying his work to games of dice and cards, Bernoulli studied such trials in *Ars Conjectandi*, published in 1713, which is considered a founding work in the field of probability.

Bernoulli's formula for computing the probability of making 11 of 13 shots, given a shooting percentage of 45.3%, is $C(13,11) \times (0.453)^{11} \times (1 - 0.453)^2$.

Let's take this piece by piece. First, $C(13,11) = 13!/(2! \times 11!) = 78$—the number of possible combinations where the order of hits

and misses does not matter. Next, $(0.453)^{11}$ represents Curry's eleven successes (with probability 0.453), while $(1 - 0.453)^2$ represents the two missed shots (with probability $1 - 0.453$). Multiplying these pieces together gives us a probability of 0.00384, or just under 0.4%.

In general, the probability of r successes out of n attempts where the probability of each success is p is

$$C(n,r) \times p^r \times (1 - p)^{(n - r)}$$

So the probability of Curry making 13 of 17 three-pointers, as in the game mentioned at the end of chapter 2, is

$$C(17,13) \times (0.453)^{13}(1 - 0.453)^4 = 0.00721.$$

The probability of the UConn women's 111-game winning streak, given a 99% chance of winning each game, is also a Bernoulli trial:

$$C(111,111)(0.99)^{111}(1 - 0.99)^0 = (0.99)^{111}$$

This is the same value we found earlier.

The highlight reel can now take on a whole new meaning. You are armed with a toolbox to analyze streaks. If the series is relatively short, like Curry's 11 of 13 shots, and the shooting or hitting percentage can be easily approximated by flipping coins or rolling dice, a quick simulation can yield insights. For lengthier streaks, like the Huskies' 111-game run, plug the numbers into Bernoulli's formula, then grab a calculator or use Google to do the computation. These tools alone can clarify whether a highlight is actually an unforgettably improbable event.

....

Analytic Toolkit

Google's virtual dice roller, introduced in chapter 3, offers the option of multiple dice. As before, type "roll dice" into Google, then click the square in the row above the "Roll" button twice to add two more six-sided dice to the one that appears by default. (The roller also offers 4-, 8-, 10-, 12-, and 20-sided dice.) Click "Roll" repeatedly and see if you can match the UConn women's basketball team's winning streak.

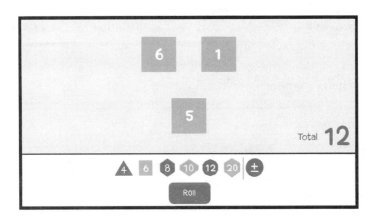

Personal Training—Workout #4

Combinations play an important role in computing probabilities. The 2019–20 National Hockey League playoffs were scheduled to begin in April, a few days after the end of the regular season. The season was suspended in March due to the COVID-19 pandemic, and in

late May, the league settled on a 24-team playoff in lieu of the usual format. The playoffs were held in two "hub" cities—Edmonton and Toronto—to reduce team travel and minimize the risk of spreading COVID-19.

Question: Possible playoff locations were limited to Canadian cities. From the seven Canadian cities with NHL arenas (Calgary, Toronto, Edmonton, Vancouver, Ottawa, Winnipeg, and Montreal), how many ways could the two hubs be selected?

Work out your answer, then check out the discussion at the back of the book.

5

UNBREAKABLE TENNIS

There have been many great rivalries in tennis history: Serena Williams versus Venus Williams, Rafael Nadal versus Roger Federer, Steffi Graf versus Monica Seles, John McEnroe versus Bjorn Borg, Martina Navratilova versus Chris Evert, Pete Sampras versus Andre Agassi. Such rivalries can stretch over decades. Tennis fans watch as the players' games evolve with age and experience. In some cases, the players themselves evolve, as exemplified by Andre Agassi's hair, which transitioned from a 1990s rocker look to clean-shaven.

A tennis match is broken into sets. The best of three or five sets, depending on the match, determines the winner. A set is broken into games. The first player to reach six or seven games, and to be

ahead by two games, wins the set. If the set reaches a 6–6 tie, a tiebreaker takes place, and whichever player reaches seven or more points with a lead of two points the set. Throughout the match, service alternates—one player serves in the first game, the other in the second, and so on—as the server has a significant advantage. For the best players, the percentage of service games won can be as high as 90%. Andre Agassi, over his career, served in 10,419 games and won 84% of them. We can simulate Agassi's winning percentage in service games by rolling a die. As long as you don't roll a six (83.3% likely), Agassi wins the game and "holds serve."

But we can also get more granular and simulate a game point by point. For each point, a ball is served; if it lands in the service box, play proceeds until the point is won. If it lands outside the service box, the server is given a second serve. That serve either lands within the service box and play proceeds, or the point is lost by the server with a "double fault." The first player to win four or more points and be ahead by two points wins the game. In a tennis game, the score begins at "love," or zero points. Winning a point takes a player to 15, then 30, then 40, and then game point, which wins the game unless the players are tied at 40–40, which is called "deuce." From deuce, the game proceeds until a player pulls ahead by two points.

To simulate Agassi's play point by point, we'll use his career service stats:

- first serves in: 41,133/65,690 = 62.62%
- first serves won: 29,998/41,133 = 72.93%
- second serves won: 13,256/24,557 = 53.98%

Think about how we might estimate these stats by flipping a coin. Then use the worksheet to play point by point in an Agassi service game.

Worksheet: Serve Like Agassi

To play a point like Agassi, follow these steps:

- First serve: flip a coin three times.
  - If exactly two tails, serve is out. Go to second serve.
  - Else, serve is in. Play proceeds: flip a coin two times.
    - If exactly two tails, Agassi loses the point.
    - Else, Agassi wins the point.
- Second serve: flip a coin one time.
  - If tails, Agassi loses the point.
  - If heads, Agassi wins the point.

How many straight service games can you win as Agassi?

The 2001 US Open quarterfinal match between Andre Agassi, seeded #2 in the tournament, and Pete Sampras, seeded #10, was a meeting of legends. Over his career, Sampras had held the world's #1 ranking for a then-record 286 weeks. Agassi had topped the list for 101 weeks.

Sampras's serve is often listed as one of the best ever in men's tennis. Agassi was known for his return. When the two met for the 2001 quarterfinal in Arthur

Ashe Stadium, they had played each other thirty-one times, with Sampras winning seventeen matches. That day's match would immediately be the talk of tennis.

The first set reached a 6–6 tie with neither player losing serve. The tiebreaker then reached 7–7 before Agassi pulled away and won 9–7, thus taking the first set 7–6. To this point in his career, Agassi had won all but one of fifty matches in which he won the first set. In his meetings with Sampras, the winner of the first set had won twenty-four of the thirty-one previous matches.

The second set also reached a 6–6 tie without a loss of serve, as did the third, and the fourth. In the end, despite Agassi's first-set momentum, Sampras won 6–7, 7–6, 7–6, 7–6. In the entire match—forty-eight straight games—neither player lost serve. Let's estimate the probability of this remarkable event.

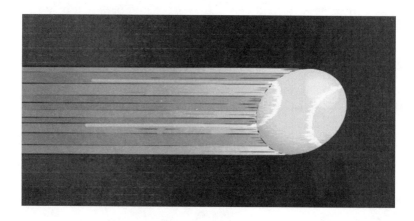

We've already tried playing as Agassi. Now, we'll also play as Sampras. His service game stats are:

- first serves in: 39,508/66,435 = 59.47%
- first serves won: 31,969/39,508 = 80.92%
- second serves won: 14,167/26,927 = 52.61%

We will assume that both players' career stats translate into probabilities for this match. They wouldn't entirely. In playing Sampras, Agassi was returning one of the best serves in tennis history. And Sampras was playing against Agassi's legendary return of a second serve. But even without exact probabilities, we can get a sense of the improbability of their unbreakable match.

To play like Sampras, we'll use a batch of random numbers between 0 and 1. How do we get them? One way is to open a Google Sheets spreadsheet and type =RAND() into 100 cells—and there they are!

Armed with randomness, here is how a computer can play like Pete Sampras:

- Pick a uniform random number between 0 and 1. If the number is less than 0.5947, Sampras's first serve goes in.
- If the first serve goes in, pick another uniform random number between 0 and 1. If it is less than 0.8092, then Sampras wins the point.
- If the first serve didn't go in, pick a uniform random number between 0 and 1 for the second serve, and if it is less than 0.5261, then Sampras wins the point.

Now that we can simulate both Agassi's and Sampras's service games, we just keep track of how many points each player has and who wins each game. If Sampras is serving and Agassi wins, then Sampras

didn't hold serve. Simulating 100,000 service games for each player, Pete Sampras holds serve 89% and Andre Agassi 84% of the time, with these percentages being the same across multiple simulations.

So what is the probability of a set reaching a 6–6 tie, with neither player losing serve? The probability Sampras wins six straight service games is $0.89^6 = 49.7\%$. Similarly, the probability Agassi wins six straight service games is $0.84^6 = 35.1\%$.

The probability of Sampras and Agassi reaching 6–6, then, is $0.497 \times 0.351 = 17.4\%$. So the probability of their playing four sets with neither player losing serve is about $0.174^4 = 0.1\%$, or about 1 in 1,000—all the more striking when you remember that the players only met thirty-five times over their entire careers.

....

Playbook

A big advantage of **computer simulation** is the speed of digital computation. It's impractical to roll a die a hundred thousand times, but a computer can quickly run ten thousand, a hundred thousand, or even a million simulations. Simulation is, by definition, based on randomness, so computations can vary. To compute the probability of an event occurring, program a computer to simulate the phenomenon. If you divide the number of times the event occurs in the simulation by the total number of simulations, you get an estimated probability for the event. How reliable is that number? A quick test is to run the experiment again. Do the estimated probabilities agree to a suitable number of decimal places? If not, increase the number of simulations. For instance, suppose you run ten thousand simulations and get 0.15 for the estimated probability of success—but then get 0.11 on the next run. Want more accuracy? Increase the number of simulations to twenty thousand, fifty thousand, or more.

This isn't the only improbable outcome in a tennis match. Consider a golden match, which occurs when a player does not lose a single point in an entire match. There are only four documented golden matches, the first attributed to Hazel Wightman in a 1910 amateur match. Wightman dominated women's tennis, winning seventeen Grand Slam titles and forty-five US titles during her career. She is known as the "Queen Mother of American Tennis" or "Lady Tennis" for her lifelong contributions to women's tennis.

By using a computer in this chapter, we could integrate a more complicated game structure, simulating a tennis match point by point. While computer simulation allows greater precision than rolling a die or flipping a coin, we saw again how those simple methods could give a sense of the improbability of an event. So, keep that die and coin handy if you decide to watch tennis. If you want to be more exact in your computations, you can turn on your computer and have it simulate a seemingly improbable game, set, or match. If you code, you can write a program to simulate the game many times. Or you can have a spreadsheet pick random numbers and play the game yourself to get a sense of whether you are watching the improbable and possibly unforgettable.

Analytic Toolkit

Again, you can type =RAND() in a cell of a Google Sheets spreadsheet to produce a random number. There are also online random number generators. Remember, our random numbers should be **uniformly distributed** over the interval 0 to 1, that is, they should be equally likely to be any number in that range. Uniform distribution is important since other options are also available with random number generators.

Personal Training—Workout #5

Let's have a personal training workout with Serena Williams, whose serve has been clocked at 128 mph (at the Australian Open in 2013).

Between 2012 and 2017, according to ESPN, Serena's percentage of aces (unreturned first serves) was 14.4%. For comparison, the ace percentage in all women's matches from 1999 to 2020 was 4.5%. Let's look at Williams's 2012 Wimbledon semifinal against Victoria Azarenka, where Serena hit an astounding 24 aces in 71 serves, according to tennisabstract.com.

Question: Assuming a 14.4% chance of serving an ace, what's the probability of Serena Williams having exactly 24 aces in 71 serves?

Approach 1: Treat this as a Bernoulli trial (described in the last chapter) and compute the probability.

Approach 2: Feeling fuzzy, or simply want to confirm your Bernoulli trial calculation? Write a computer simulation to estimate the probability. Explain the steps in the program.

See what you come up with, then check out the discussion in the back of the book.

6

IMPROBABLE NFL PLAYOFF

The 2019 National Football League AFC wild-card playoff game started off one-sided. A touchdown and two field goals had the Buffalo Bills up 13–0 at halftime, and another field goal in the third quarter made it 16–0. Then came the rollercoaster. The Houston Texans scored nineteen straight points. Buffalo seemed set to attempt a tying field goal, but didn't. Houston only needed to pick up a first down to win, but didn't. The Bills had one last chance for one last drive and, with no timeouts and seconds on the clock, evened the score with a field goal, sending the game to overtime.

Announcers marveled. A score of 19 is unlikely, a 19–19 tie all the more improbable. How improbable? To answer that, we first need to consider how a team might reach a score of 19 points. Some combinations seem less remarkable than others. The Bills, for instance,

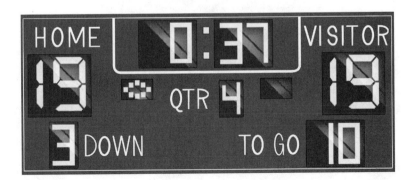

scored a touchdown and extra point, followed by four field goals. The Texans, by contrast, took a less likely route: a touchdown with a two-point conversion, then a field goal, then another touchdown with a two-point conversion.

There are eighteen possible ways to score 19 points in the NFL, some of which would only result from bizarre play on the field. Among the scoring combinations:

- 3 + 3 + 6 + 7: two field goals, two touchdowns with only one extra point
- 7 + 7 + 3 + 2: two touchdowns with extra points, a field goal and a safety
- 2 + 2 + 2 + 2 + 2 + 2 + 2 + 2 + 3: eight safeties and a field goal

Can you work out the other thirteen?

How much more likely are some of these scoring scenarios than others? Let's next look at how often different scoring events occurred in the 2019–20 NFL season.

Scoring event	Occurrences
Touchdown (6 points)	1,244
Extra point (1,210 attempted)	1,136
Two-point conversion (113 attempted)	54
Field goal (3 points)	802
Safety (2 points)	17
Total drives	5,728

(This list includes only rushing and reception touchdowns, not kickoff and punt returns and turnovers returned for a touchdown.) Given the frequency of each scoring event and the total number of possessions, we can compute some probabilities.

- touchdown (TD): 1244/5728 = 0.217
- extra point after TD: 1136/1210 = 0.934
- two-point conversion after TD: 54/113 = 0.478
- field goal: 802/5728 = 0.140
- safety: 17/5728 = 0.00297

We can approximate these probabilities by rolling a die.

....

Worksheet: NFL Drive

Roll a die four times to determine the outcome of your NFL possession:

- Safety—if you roll four 2's, four 3's, four 4's, or four 5's, giving a probability of 4/1296 = 0.003.
- Touchdown—if you roll
    - ° 1 on the first roll, giving $6^3 = 216$ options or
    - ° 2-3 or 3-4 on the first two rolls giving an additional 2 ×

$6^2 = 72$ options, so the probability of a touchdown is $(216 + 72)/6^4 = 0.22)$.

- Field goal—if you roll 6 on the first roll, giving a probability of $6^3/6^4 = 0.17$.

When you score a touchdown, you get 6 points. Determine if you add 0, 1, or 2 additional points by rolling a die twice. Note that two-point conversions are attempted after a touchdown $113/(113 + 1210)$, or 8.5% of the time.

- If you roll 1-1, 2-2, or 3-3, you are attempting a two-point conversion, which equates to a probability of $(3 \times 6^2)/6^4$, or 8.3%. Roll one die a third time, and if you get 1, 2, or 3, you score 2 points, which occurs 50% of the time!
- Otherwise (if you didn't roll 1-1, 2-2, or 3-3) you are attempting an extra point. Roll one die two more times. If you roll anything but 1-1 or 2-2, you score 1 point, which occurs with a probability equal to $1 - (2 \times 6^2)/6^4 = 0.944$.

NFL teams average about twelve offensive possessions per game. So, using one die, you can simulate the scoring in an entire game by playing twenty-four possessions—twelve for each team—according to the rules above. Just tally up the points to compute the final score.

How often do you get 19? Which scores are most common? To determine this, we need to play a game many times. So let's turn again to the computer. We'll play over a million games and keep track of how often 19 points are scored by one of the teams, and in what combination.

How do we get a computer to simulate this type of game? Let's simplify things and only consider our earlier probabilities for a touchdown (0.217), field goal (0.140), or safety (0.00297). The computer will again pick random numbers that are equally likely to be any number between 0 and 1. Here are the steps for the computer:

- Pick a uniform random number between 0 and 1. If the number is less than 0.217, a touchdown is scored. Then, pick another uniform random number.
  - If the number is less than 0.085, a two-point conversion is attempted. Pick another uniform random number, and if it is less than 0.478, the two-point conversion succeeds.
  - Else (if the number is greater than or equal to 0.085), an extra point is attempted. Pick another uniform random number, and if it is less than 0.934, an extra point is scored.
- If the number is greater than or equal to 0.217 and less than $0.217 + 0.140 = 0.357$, a field goal is scored.
- If the number is greater than or equal to 0.357 and less than $0.217 + 0.140 + 0.00297 = 0.360$, a safety is scored on your possession giving the other team 2 points.
- Else, the possession doesn't end in a touchdown, field goal, or safety.

When we have a computer play many games following this procedure, with twelve possessions per team, and tabulate the final

scores, we find that scores like 20 or 17 are more than twice as likely as scores of 19. There is only a 2.9% chance of scoring 19 points. The most likely way to get to 19 points is two field goals and two touchdowns with only one extra point. Not far behind is the Bills' combination: one touchdown with an extra point plus four field goals. The chance of getting two touchdowns, both with two-point conversions, and a field goal is only 0.01%. We're about a hundred times more likely to reach 19 points the way the Bills did than the way the Texans scored. Consider the game's mix of improbable and even more improbable events, and it becomes all the more memorable.

Oh yes, how did the game turn out? In overtime, Houston quarterback Deshaun Watson was sandwiched between rushers. He bounced off one, then another, before throwing a short pass to Taiwan Jones, who darted down the field for a huge gain. The next play, following two timeouts, was a successful field goal. The game ended 22–19, with the Texans' kicker being hoisted into the air.

When will the next seemingly improbable game occur? When it does, start rolling a die or run some computer code to see just how improbable it is.

Analytic Toolkit

Where do all the stats come from? How do we know the number of drives, touchdowns, field goals, and safeties in a season? The collection of databases at www.sports-reference.com boasts a huge array of stats on professional and college sports. I looked up the NFL's 2019–20 season (at www.pro-football-reference.com/years/2019/), then searched for stats on drives and scoring offense.

How different are other years' statistics? What if you combine years? Dig in and develop your own analyses!

Personal Training—Workout #6

We developed a way to simulate a game of football by rolling dice. Let's take two steps toward simulating baseball in the same way.

In the 2016 Major League Baseball season (according to baseball-reference.com), there were 165,561 at-bats resulting in 27,539 singles, 8,254 doubles, 873 triples, and 5,610 home runs. We can convert each to a probability and link this to some roll or combination of rolls of a die. For each at-bat in 2016, there was a 27,539/165,561 = 0.166 probability of hitting a single. We can simulate this by rolling one die. If you roll a 1, you hit a single, since 1/6 = 0.166. The chances of hitting a double (8,254/165,561 = 0.050) are similar to those of first rolling a 2 and then rolling either a 1 or a 2 (2/36 = 0.055).

Question: What roll or combination of rolls would you choose to approximate a home run in a simulation where the roll of three dice determines the outcome of an at-bat?

Decide on your optimal roll or combination of rolls, then check out the discussion at the back of the book.

7

FIFA OCTOPUS ORACLE

The most popular sports around the world vary—cricket may fill screens in India while sumo streams in Japan. Still, some sporting events are worldwide phenomena, and few if any are bigger than the FIFA World Cup. A combined 3.6 billion viewers—more than half the global population aged four and over—tuned in to the 2018 competition.

Along with the scores, tackles, yellow and red cards, and penalty kicks come the inevitable predictions. Who will win each match? Who will make it out of the group stage? Will a perennial powerhouse chalk up another championship, or will an underdog make

history? In the 2010 men's World Cup, Spain gained bragging rights with a 1–0 victory in the final over the Netherlands. But not all the stars were on the pitch. In fact, one was not even on dry land.

Paul the Octopus, who lived at the Sea Life Center in Oberhausen, Germany, established himself as an animal oracle. As the tournament progressed, his uncanny knack for predicting winners caught worldwide attention.

He correctly predicted all seven of Germany's games, including its semifinal loss. He also predicted Spain's win over the Netherlands in the final.

How did Paul pick his winners? To divine the result of an upcoming match, his keepers would present him with two boxes, each containing a mussel. The boxes were identical except for the flags that adorned them. The predicted winner was the team whose flag decorated the box Paul ate from first.

Was the seeming accuracy of Paul's selections simply luck? Let's look at his string of correct predictions for Germany's seven games. We'll start by assuming his picks were random, equivalent to flipping a coin. Flip a coin seven times. Do you get all heads? We've seen this type of analysis. The chances of flipping seven consecutive heads are $(1/2)^7$, which is 1 in 128, about 0.78%. Could guesswork

achieve that? We'll return to this question in a moment.

Over Paul's two-and-a-half-year life, he mainly predicted the outcomes of the German national team's international matches. At his death, Paul had a success rate of 85.7%, having predicted twelve of fourteen matches.

His correct prediction for the 2010 final made Paul an instant hero in Spain. The Spanish team held the trophy high after their win. The next day, the aquarium presented Paul with his own (fake) World Cup trophy, containing mussels—and a Spanish town made him an honorary citizen.

Paul's success was certainly memorable, but is his celebrated clairvoyance unforgettable? Let's compute the probability of matching Paul's performance with coin flips.

As with his predictions of Germany's fate in the 2010 World Cup, we can assess his lifetime record with coin flipping. Flip a coin fourteen times and see if you get twelve heads. As always, you could repeat those fourteen flips a million times and tabulate the likelihood of getting twelve heads. But instead, let's calculate it. Remember the Bernoulli trial from chapter 4? Using that formula, our chances here become $C(14,12) \times p^{12} \times (1 - p)^2 = ((14 \times 13)/2 \times 1) \times (0.5)^{12} \times (1 - 0.5)^2 = 91 \times (0.5)^{14} = 0.00555$.

If happenstance led Paul to his successful guesses, we would expect roughly equal numbers of hits and misses. He would have been just as unlikely to make only two correct predictions as to miss only two. Making fewer than two or more than twelve correct predictions would be even less likely. If Paul were guessing, what's

the probability that he could generate 0, 1, 2, 12, 13, or 14 correct predictions from fourteen random guesses?

To restate this question: What's the probability that random chance (a coin flip) will generate results equal to or rarer than Paul's guesses?

....

Worksheet: Paul the Octopus's Predictions

Let's make FIFA predictions by guessing.

1 Flip a coin 14 times.
2 Keep track of the number of heads and tails in the grid below.
3 You match Paul's unlikeliness by getting heads 2 or 12 times.
4 You beat Paul's unlikeliness by getting heads 0, 1, 13, or 14 times.

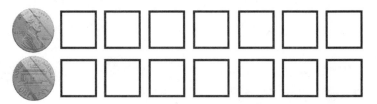

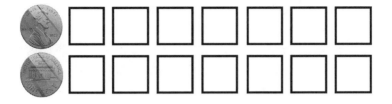

Just as before, the Bernoulli trial allows us to compute these probabilities exactly. Making 12, 13, or 14 correct guesses has a likelihood of $(14 \times 13)/(2 \times 1) \times (0.5)^{14} + 14 \times (0.5)^{14} + (0.5)^{14} = 0.00647$, or 0.65%. Making 12, 13, or 14 incorrect guesses is exactly as likely, since the probability of being right (heads) is the same as that of being wrong (tails) when you flip a coin. So this probability, too, is also about 0.65%.

The overall probability, then, of guessing right or guessing wrong two of fewer times out of fourteen is 1.3%. That's about 1 in 75. Some argue that this may be mildly memorable but isn't especially improbable. That's not just anti-octopedal naysaying. Given the worldwide notoriety of the World Cup and the number of different ways people try to divine the outcomes, someone or something will have a higher than expected success rate, and Paul could simply be the one who got lucky.

In statistics our result of 1.3% is called a *p*-value. In fact, it's a strong *p*-value, which could make someone sit up and take notice. It's not compelling evidence for octopus ESP, just sufficient reason to seek to replicate the result. Could Paul predict another FIFA World Cup as accurately? We'll never know. So the source of Paul's predictive power remains an unanswered question.

Analytic Toolkit

A *p*-value is a measure of the probability that an observed difference could be the result solely of random chance. The lower the *p*-value, the greater the **statistical significance**. A *p*-value of less than 0.05 (5%) is often the threshold for calling something statistically significant. Depending on the application, a *p*-value of 0.01 may be required.

Personal Training—Workout #7

In 2015, President Barack Obama correctly predicted Connecticut as the women's March Madness champions. He didn't fare as well on the men's side of the tournament, which was noticed by Emilia, an eleven-year-old North Carolinian, before the 2016 tournament. "You are a great president, just not the best bracket picker," she told him in a letter.

Obama's 2016 bracket predicted the 13-seed University of Hawaii's major upset of 4-seed University of California, Berkeley. Keep in mind, Obama grew up in Hawaii. Overall, Obama correctly picked 23 of 32 games in the first round of the 2016 men's tournament. Could he have just been guessing? Our analysis of Paul's picks can help us answer.

Question: What's the probability that random chance (a coin flip) would generate results at least as good as Barack Obama's 23-of-32 first round 2016 March Madness picks?

Work on your answer, then turn to the back of the book for a discussion of this question.

8

SUPER-SIZED SUPER BOWL TD

In late January 1986, the Chicago Bears met the New England Patriots for Super Bowl XX in the New Orleans Superdome.

The Bears were coached by Mike Ditka, who had played tight end for the Bears during their last trip to the championship game, which they won in 1963, before the advent of the Super Bowl.

The 1985–86 Bears had lost only one of their eighteen regular and postseason games that year. By contrast, the Patriots had lost five, and hadn't won their division.

The game started well for the Patriots, who took the quickest lead in Super Bowl history (at the time) when Tony Franklin kicked a 36-yard field goal only a minute and 19 seconds into the game. The glimmer faded quickly, though, as the Bears built a 23–3 lead by halftime. At that point, Chicago had gained a total of 236 yards versus New England's –19.

The outcome almost seemed predestined. The Bears had posted the best regular season record at 15–1. Some of the players had recorded a

rap song, "The Super Bowl Shuffle," the day after their only loss of the season and two months prior to the championship game. The single sold half a million copies; in February it would peak at No. 41 on the Billboard Hot 100 chart, and it was even nominated for a Grammy.

The telecast of the game was watched by an estimated 92.6 million viewers. Bears defensive end Richard Dent had 1.5 quarterback sacks, two forced fumbles, and one blocked pass, and was named the game's Most Valuable Player.

The moment that will be the focus of our analysis came in the

THE SUPER BOWL SHUFFLE
THE CHICAGO BEARS SHUFFLIN' CREW

third quarter when 300-pound-plus defensive tackle William "the Refrigerator" Perry trotted onto the field and lined up as a running back. The ball was snapped, and Bears quarterback Jim McMahon handed the ball to the Fridge, who hurled his body forward for a one-yard touchdown, spiked the ball, and cemented a place in the Super Bowl highlight reels and in popular culture.

The game ended with the Bears winning 46–10, setting a new NFL record for margin of victory. The Bears also set or tied Super Bowl records for sacks (7) and fewest rushing yards allowed (7).

The team had many notables. Ditka was named NFL Coach of the Year by AP, Sporting News, and UPI. Mike Singletary was the NFL's Defensive Player of the Year. The defense is frequently cited as one of the best in NFL history. Many elements of Super Bowl XX could be analyzed. We'll examine that one-yard rushing touchdown.

Before the game, the most probable pick for a Bears rushing touchdown was future Hall of Fame running back Walter Payton—the season's UPI NFC Player of the Year. Payton also held records for career rushing yards, touchdowns, carries, yards from scrimmage, and all-purpose yards.

Perry's rushing touchdown extended a very different record. In a Monday-night game during the regular season, the Bears had needed one yard to score a touchdown and break a second-quarter tie. As he would at the Super Bowl, McMahon handed the ball to Perry, who rumbled into the endzone, becoming the heaviest man in NFL history to score a running touchdown off a set play.

Perry's touchdowns prompted significant fanfare. Were they unforgettably improbable or merely a by-product of the Bears' accolades and pop culture status? Let's dig into the numbers.

Among players who have scored Super Bowl rushing touchdowns (as of 2020), Perry is the heaviest. That season, he was listed at 325 pounds—125 pounds more than Payton and 60 pounds more than the next heaviest player to score a Super Bowl rushing touchdown. The Fridge is clearly an outlier within that select group.

But Perry was also a defensive tackle. How heavy was he for that position? If another defensive tackle launches into an endzone, should we expect a Fridge-like weight?

If we look at weights of some three thousand NFL defensive ends and tackles drafted between 1938 and 2013, we find that 97.74% these players weighed less than 325 pounds, putting Perry near the top in weight even for his position. We can plot these numbers in a histogram, which resembles a bar chart but groups quantitative data into ranges. In our histogram showing the distribution of end and tackle weights, the highest proportion of weights are between about 250 and 268 pounds (the value on the x-axis is the integer nearest the midpoint of the associated bar). The width of each bar is about nine pounds.

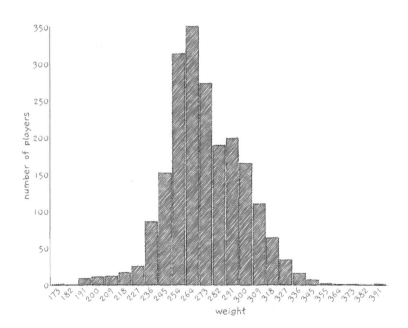

Our histogram's shape is similar to a normal distribution (or bell curve). A normal distribution represents a continuous probability, as opposed to a set of discrete probabilities like those for the possible outcomes of rolling a die. In the bell curve shown here, the dotted line occurs at the average value (also known as the mean) of the dataset. Some normal distributions are flatter than others. This is captured in the standard deviation, which represents how much data varies. The bigger the standard deviation the more the underlying data varies and the flatter the bell curve. For data following a normal distribution, you can find the percentage of data between two values, or less than a given value, if you know the mean and standard deviation.

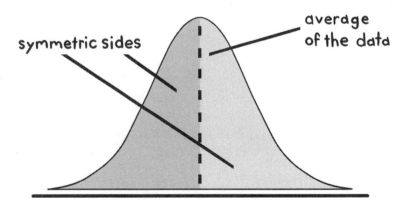

For a normal curve, approximately 68% of the data lies within one standard deviation of the mean and 95% of the data lies within two standard deviations of the mean. Our NFL defensive end and tackle weights have an average of 272.42 pounds and a standard deviation of 25.98; Perry's weight is $(325 - 272.42)/25.98 = 2.02$ standard deviations above the average. Note that Perry's weight is on the far right side of the curve. Half of the weights more than 2.02 standard deviations from the mean are on the right side (heavier than Perry) and half are on the far left (more than a hundred pounds lighter than Perry), so approximately 97.7% of NFL defensive ends and tackles are lighter than Perry, assuming a normal distribution of the data. In fact, exactly 97.3% of the data points were less than Perry's weight, demonstrating the quality of our estimate.

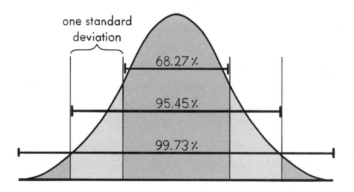

Rolling dice can approximate an event with a normal distribution. The probabilities of rolling sums from 2 to 12 with two six-sided dice are easy to compute. For example, there are six ways to roll a sum of 7 (1-6, 2-5, 3-4, 4-3, 5-2, 6-1) out of thirty-six possible rolls (6 × 6 = 36), so the central bar on the histogram is the highest, representing a probability of 6/36 = 0.16. There is only one way to roll a sum of 12: 6-6. So the probability of rolling a sum of 12 with two dice is 1/36; thus, the probability of rolling a sum less than 12 is $1 - 1/36$, or 0.972 (97.2%)—close to the chance that a random player pulled from our data will weigh at least 325 pounds.

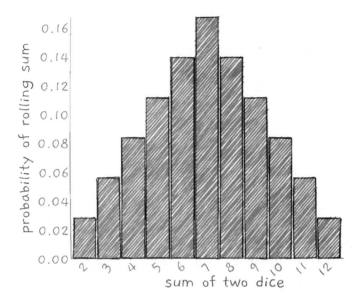

....

Worksheet: Weight and NFL Defensive Tackles

Each roll simulates pulling a random player who has played defensive line or defensive tackle in the NFL.

1 Roll two dice (or one die twice).
2 If you roll a 12 (two 6's), then your player weighs at least 325 pounds.
3 Else, your player weighs less than the Fridge.

Try it several times. Did you get anyone heavier than Perry? If so, how many?

....

In the other chapters' worksheets, we estimate an underlying probability. Here, the unlikeliness of a player of Perry's weight rushing for a Super Bowl touchdown is evident. Perry is an outlier in weight even for his defensive position. Further, he ran the ball, which is uncommon for that position. And not only did he run the ball, he scored a touchdown in the Super Bowl. All of this underscores why Perry's one-yard rushing touchdown launched him into Super Bowl lore.

Analytic Toolkit

Our histogram of the weights of defensive linemen was similar to a normal distribution but not a perfect match. This is to be expected. Let's create a histogram showing the results of rolling a die twelve times.

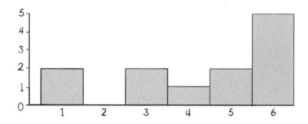

You'll notice that I didn't even roll a 2. The sample size is too small to know if we have a fair die. So let's roll it ten thousand times (a sample three times larger than our weight dataset). The new histogram still does not show exactly a 1/6 probability for each possible outcome due to randomness, but the distribution is very close to even.

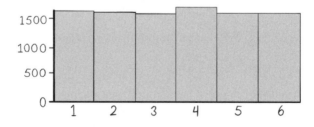

Personal Training—Workout #8

Kareem Abdul-Jabbar, who played in the National Basketball Association from 1969 to 1989, stood 7 feet, 2 inches tall. Yet he wasn't the tallest NBA player. Manute Bol, who played from 1985 to 1995, was a towering 7 feet, 7 inches. If we plot NBA players over sixty-seven seasons beginning in 1950, we find a fairly normal curve of heights. While Kareem might not be the tallest, let's get a handle on how tall he is in terms of the data.

Question: The mean height of our NBA player dataset is 6 feet, 6 inches, with a standard deviation of 3.65 inches. How many standard deviations from the mean is Kareem Abdul-Jabbar's height?

Calculate your answer, then turn to the back of the book for a discussion of this question.

9

SCORING CONFIDENCE

In April 2019, a crowd of over seventeen thousand watched Alex Morgan race toward the goal, sidestep an Australian defender, and, with a sliding kick, place the ball into the net. The spectators erupted, knowing they had just witnessed the American soccer star's one hundredth career goal in international competition.

Morgan's run to this milestone culminated in a scoring streak of twenty-seven goals in thirty-two games that began in 2017. Anticipation built as she scored goal 98 against Scotland in November 2018, but the next two goals required patience. Goal 99 came more than three months later, at the end of February 2019, against Japan. in late February 2019. Finally, some five weeks later, Morgan hit the century mark.

Morgan's scoring skill had been evident early in her career. At the 2011 World Cup, the twenty-two-year-old—the youngest player on the US roster—scored in the semifinal against France and again in the final against Japan. She would be the third youngest player in the history of US women's national soccer to score 100 goals (at 29 years, 276 days). Only Mia Hamm (26 years, 185 days) and Abby Wambach (29 years, 47 days) had been younger. She was also the fourth fastest: only Wambach (129 caps), Michelle Akers (130), and Hamm (156) had needed fewer games.

Let's convert these stats to goals per cap. Morgan scored 100 goals in 159 international games, for an average rate of $100/159 = 0.629$ goals/cap. (The term "cap" comes from the time when players received actual caps for international play.) Wambach averaged 0.775 goals/cap; Akers, 0.769 goals/cap; and Hamm, 0.641 goals/cap.

Game-to-game performance fluctuates, of course. There were

months in which she scored no goals, and games in which she scored two or three (a complication, discussed later, that we will ignore in our first simulations). Can we nonetheless use Morgan's stats to get a sense of her underlying ability?

Let's simulate Alex Morgan's play with the roll of a die. In chapter 2, we gave ourselves a three-point shooting percentage slightly better than Steph Curry's. Here, your goal-per-cap rate will be slightly higher than Alex Morgan's. Her average (for those first 100 goals) was 0.629; yours will be 0.667 goals/cap. Each roll of a six-sided die will represent one game; if you roll a 1, 2, 3, or 4, you score. We'll start with just ten games.

Worksheet: Scoring Like Alex Morgan

To see if you score a goal in a cap:

1 Roll a die.
2 If you roll a 1, 2, 3, or 4, you score; else, you don't.
3 Repeat this ten times and keep track of your progress with the chart below.

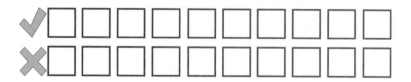

In how many games did you score? With only ten games, I'm not very confident that I can predict your results. I can say it's fairly likely you scored in six, seven, or eight games. How likely? If we use a computer to repeat our ten-game series ten thousand times, we find that 67% of the time, we score in six, seven, or eight games (for a goal/cap rate between 0.6 and 0.8). And about 95% of time, we score in four to nine games (a goal/cap between 0.4 and 0.9). If we want to be 95% sure of the underlying goal/cap rate for a player who played only ten games, we have to allow a pretty big spread of potential values!

For a player with a longer career, however, we can be more certain. This time, let's roll a die a hundred times and record the number of games in which we score a goal (as before, by rolling a 1, 2, 3, or 4). Once again, we'll do this ten thousand times. This time, you'll score in 58 to 76 games about 95% of the time. In other words, I'm about 95% confident that your goals/cap rate is between 0.58 and 0.76.

You could roll a die to

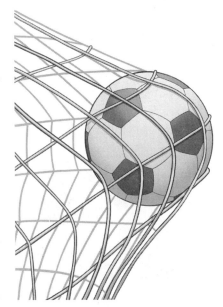

play a hundred games ten thousand times by hand, but it's certainly faster to simulate those rolls using a computer. With a computer we can also set aside our virtual die and use Alex Morgan's actual goal/cap rate of 0.629. Here's our automated procedure:

- For each game, pick a uniform random number between 0 and 1.
- If the number is less than 0.629, Morgan scores in the game. Else, she doesn't.
- Play 159 games, tally the number of games in which she scores, and compute the goals/cap rate.
- Repeat ten thousand times.

When the computer plays ten thousand sets of 159 games, Morgan's goals/cap rate falls between 0.55 and 0.71 for about 95% of our simulations. This is called a 95% confidence interval. If Morgan had

played ten times as many games, then the spread in the confidence interval would be smaller.

Alex Morgan will continue to score as she plays more games. At the end of her career, we shouldn't expect her goals/cap rate to be exactly 100/159, but we can be fairly confident that it will be between 0.55 and 0.71.

Playbook

We have been calculating and using Alex Morgan's goals/cap rate as if her scoring were entirely the result of individual action. This overlooks the team aspect of soccer: Morgan's goals are connected, for instance, to her teammates' passing and positioning on the pitch—a collective effort to help her specifically (not all players equally) score goals. Another wave of analysis might work to create or analyze a stat that more fully integrates these elements of the game. Simplifying assumptions are natural and necessary in analytics and modeling. An important part of the work is remaining aware of the assumptions you are making so you can recognize the limits they impose and their effects on the applicability of your analysis.

One use of confidence intervals is to compare players. Suppose we run the same computer simulation using statistics for Abby Wambach, who scored 100 goals in 129 games (0.775 goals/cap). For Wambach, the computer plays 129 games ten thousand times, and we find we can be 95% confident her goals/cap rate falls between 0.70 and 0.85. Given these numbers, for Wambach and Morgan to have the same underlying scoring ability as they reached a hundred goals, Wambach would have to be performing at the low end of her confidence interval and Morgan at the high end of hers.

Want to get even more realistic? You could add a variable to your simulation to allow for more than one goal per game, which connects to how often the player scored. An advantage of computer simulation is how easily such factors can be included.

Players develop and change over time. How will Alex Morgan play after her first hundred goals? Time will continue to tell, of course, and during that time, we'll likely have much to cheer for!

Analytic Toolkit

We assumed Alex Morgan's first hundred goals were evenly distributed across the international games she played up to April 2019. In fact, Morgan scored in 70 of her first 159 caps. Of those 70, she scored one goal in 44 games, two goals in 22 games, and three goals in 4 games. We can approximate the likelihood of these possibilities by rolling two six-sided dice. On your first roll, let sixteen possible outcomes equate to Morgan scoring in a game (16/36 = 0.444; 70/159 = 0.440). If she scores, roll the two dice again; let two outcomes equate to Morgan scoring three goals in the game, eleven outcomes to her scoring two goals, and the remaining twenty-three outcomes to her scoring one goal. Sports analytics often involves adding complexity to models in this incremental way. Try it and play even more like Alex Morgan.

....

Personal Training—Workout #9

Abby Wambach is among the leading career scorers in international soccer. From 2001 to 2015, in 256 caps, she scored 184 goals, giving her a goals/cap rate of 0.719 (near the low end of the confidence interval we calculated based on her first 100 goals). As with our partial-career simulations for her and Alex Morgan, we can use a computer to simulate 256 games with Wambach's career goals/cap rate.

Question: What would be the steps a computer would follow to find Wambach's goals/cap rate over 95% of the simulations?

Write down your steps, then turn to the back of the book for a discussion of this question. If you program, you may want to write your own simulation.

TIGER'S CONSISTENCY

We've looked at a variety of statistically unlikely streaks. In chapter 4, we analyzed the UConn Huskies' three-season run of 111 straight wins. And in chapter 3, we considered Joe DiMaggio's 56-game major-league hitting streak. (He'd also had a 61-game hitting streak in the minor leagues.) After the hitless game that ended his MLB hitting streak, he hit in the next 16 games—meaning he hit in 72 out of 73 games. Some of the best athletes and teams in sports history sustain such remarkable streaks due to both ability and consistency.

In the previous chapter, we assessed Alex Morgan's underlying ability. Here, we'll analyze a streak, in golf, through the additional lens of consistency.

As pro golfer Arnold Palmer has said, "Golf is deceptively simple and endlessly complicated." In most PGA Tour events, after two rounds roughly half the players make "the cut" and compete in the final rounds of the tournament, playing for prize money and a championship. Our analysis will focus on Tiger Woods, who between 1998 and 2005, made the cut in 142 consecutive tournaments, a PGA record. Only two other modern players even have streaks that hit triple digits, Byron Nelson with 113 and Jack Nicklaus with 105. During Woods's streak, he won thirty-six tournaments, including eight majors.

Suppose we model Tiger's dominance by assuming he was 95% likely to make the cut; that would give him a $0.95^{142} = 0.07\%$ chance of attaining his streak. Let's dig into the numbers a bit more and integrate his consistency.

In the PGA, the average first round score is about 70 with a standard deviation of about 4. Let's assume golfers make the cut if their two-round score is at most 140. We'll model a round in the PGA by rolling two dice. You score 70 with a two-dice roll that sums to 7. A sum of 6 corresponds to scoring one stroke below 70 (69), and a sum of 8, one stroke above (71). For each number below a sum of 6, we'll subtract two additional strokes, and for each number above a sum of 8, we'll add two strokes. So a sum of 10 yields a one-round score of $70 + 1 + 2 + 2 = 75$. A sum of 2 corresponds to a spectacu-

lar score of 61. How well does this dice model approximate our un-derlying model of a mean of 70 and standard deviation of 4? Recall that about 68% of the data should lie within one standard devia-tion of the mean. In our dice model, rolls that sum between 5 and 9 are within four strokes (1 standard deviation) of 70 (the mean), equating to 24/36 of the possible rolls, or 66.6% of the data. Our dice model is thus a fairly good approximation. Remember, the cut is based on a two-round score. You'll make the cut if two two-dice rolls (or one four-dice roll) sums to at most 14.

Worksheet: Play Like a PGA Golfer

Can you make the cut in thirty straight tournaments?

1 Roll four dice (or a single die four times or two dice twice) and calculate the sum.

2 If the sum is 14 or less, you make the cut. Mark a box in the grid below.

3 If the sum is 15 or more, you don't make the cut, and your streak ends.

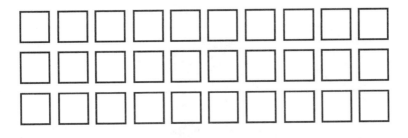

It will help us in the coming pages to directly compute the probability of making the cut via dice rolls. There are six possible outcomes when rolling a single die, giving us $6^4 = 1{,}296$ ways to roll four dice. How many of those combinations sum to at most 14? You could tediously list every option and tabulate the answer—or take a shortcut and type "number of ways to roll 4 dice with sum <= 14" into WolframAlpha.com. Either way, you'll find there are 721 ways to sum to 14 or less when rolling four dice. (See the Analytics Toolkit at

the end of the chapter for details on using WolframAlpha to count possible combinations.)

The probability of a four-dice roll summing to at most 14, then, is 721/1296, giving you about a 55% chance of making the cut in a given tournament. Your probability of making the cut in thirty tournaments in a row, then, is $(721/1296)^{30}$, or about 0.00000002. Matching Tiger's record—making the cut 142 straight times—is, using this model, amazingly improbable: 7×10^{-37}, or about a 1 in 1.5 $\times 10^{36}$ chance! Were Tiger's chances really that slim? Remember, he wasn't an average PGA player. To get a more realistic probability, we'll need to fold in Tiger's ability relative to the field.

For this, we turn to the strokes-gained statistic, which was formally implemented by the PGA Tour for the 2004 season. The idea behind strokes gained is easy enough. We have collected a great deal of data from actual play by professional golfers and know the average score for a given hole for every location from which the ball might be played. Strokes gained for a shot equals the average score for the hole based on the position of the ball before a shot minus the average score for its position after the shot. Suppose a par 4 hole has an average score of 4.2. A golfer smacks a drive far down the middle of the fairway, which lands at a spot for which the average score is 3.8. Strokes gained for the drive is $4.2 - 3.8 = +0.4$.

Strokes gained illuminates Tiger's dominance in golf. Woods is the only golfer to finish a season with an average of over three strokes gained per round—a feat he accomplished three times. Only one other golfer has had even one season with over 2.5 strokes gained per round. Tiger's average finish on the season-long strokes-gained list is 2.33.

We'll account for this in our model by assuming Tiger Woods gains two strokes per round. For now, he will still have the same standard deviation (4) as the rest of the PGA field. With this factored in, what's the probability of his making the cut 142 times in a

row? With two strokes gained per round, Tiger Woods now makes the cut, in our model, when a four-dice roll sums to 16 or less. There are 986 ways to roll four dice that sum to 16 or less. So his probability of making the cut increases to 986/1296, or about 76%. We are now about twenty quintillion times more likely to match Tiger's 142-tournament streak than we were with our earlier 55% chance. Even so, the probability is extraordinarily slim (10^{-17}).

The improved odds show the impact of ability. Now, let's fold consistency into the model. Suppose the standard deviation is much smaller for Tiger Woods than for the overall field. If we reduce his standard deviation from 4 to 2, how does this impact his chances? As before, we'll approximate Tiger's play in a round with two dice. Rolling a 7 gives Tiger a 68 for the round, thanks to his two strokes gained. Cutting the standard deviation in half means our simulated Tiger will be much more consistent. We will add only one stroke for each number above a sum of 7, and subtract only one stroke for each number below a sum of 7. So Woods scores 69 with a roll that sums to 8, 66 for a sum of 5, and so on. A four-dice roll of 18 (two rounds of 70) or less makes the cut. For this model, Wood's probability of making the cut is 1170/1296, or about 90%. Accounting for strokes gained and consistency of play in our model makes Tiger effectively three strokes per round better than the average golfer. The probability of Tiger's streak rises by a further factor of over 35 billion—but still only to about 1 in 2 million. Tiger's streak remains spectacularly improbable.

A graph showing the underlying probabilities of the possible outcomes of rolling four dice may be helpful. In our model, an average PGA golfer made the cut when we rolled a sum of 14 or less. The graph emphasizes how large a probability that leaves for not making the cut, with rolls summing to at least 15; an average PGA golfer's chance of matching Tiger's streak is minuscule, just 1 in 10^{36}. In our Tiger-specific model, our simulated Tiger made the cut with a roll of 18 or less. Note how many fewer possibilities there are for failing rolls summing to 19 or more. Tiger's streak was still improbable, 1 in 2 million, but much more possible. The modifications we made, factoring in Tiger's high strokes-gained average and reduced standard deviation, illustrate how streaks may result not simply from a player's or team's great skill but also from their consistency.

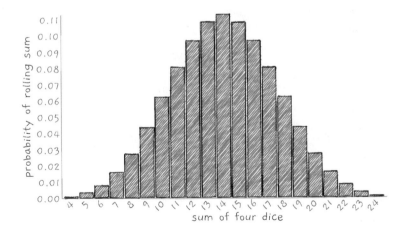

....

Analytic Toolkit

WolframAlpha.com, an "answer-engine" developed by Wolfram Research, allows you to ask mathematical (and other) questions in "natural language." You can learn the "probability of rolling sum of at most 14 with 4 dice" just by typing in that phrase (or "number of ways to roll 4 dice with sum <= 14" or some other variation). The response will spell out how your question was interpreted and give the probability of the occurrence as 0.5563, or about 1 in 1.8. Click the "Exact form" button, and this changes to 721/1296. As we have noted, 1,296 is the number of ways four dice can be rolled. Of these, 721 ways produce a sum of 14 or less. (WolframAlpha is among the services Siri uses to answer questions, which means Siri can also help you with math, including calculus!)

....

Personal Training—Workout #10

In this chapter's worksheet, our average PGA golfer made the cut when a four-dice roll summed to at most 14. We found the probability of making the cut in thirty straight tournaments to be 0.00000002. Let's switch to our final, Tiger Woods model, which assumed two strokes gained per round and a standard deviation of two strokes per round (half the PGA average). With these modifications, Tiger made the cut when a four-dice roll summed to at most 18.

Question: What's the probablity, using this model, that Tiger makes the cut in thirty straight tournaments?

Try it yourself, then check out the discussion in the back of the book.

11

MONEYBALL ANALYTICS

So far, we've focused on athletes whose efforts created historic moments. In this chapter, we'll meet a historic figure in baseball whose work was done off the field. Bill James wielded a pencil rather than a bat and created calculations that could stretch a baseball team's dollar. You can follow in his footsteps—all you need is a calculator, some data, and an appreciation of his insights.

Have you ever looked at an online sale and been wowed by the price? Analytics can sometimes create the equivalent of a 20% off coupon for, say, a player trade. In 2002 the Oakland A's had one of the lowest team payrolls in Major League Baseball—barely a third of the salary total for the chart-topping New York Yankees. The A's wanted to stretch their dollars. They needed more for their buck.

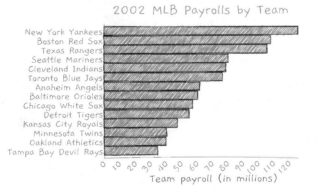

Enter Bill James. James worked nights as a security guard at the Stokely–Van Camp's pork and beans cannery. In the silence of the night, he also wrote pieces on baseball. He'd pose a lively question and answer with data and analysis. While his writing was engaging, it was unusual for the time, making it hard to find a publisher. So in 1977, he self-published what would become an annual title: *The Bill James Baseball Abstract*.

Seventy-five people purchased that first volume. Fast-forward to 2006, when *Time* magazine named Bill James one of the most influential people in the world. Why? His baseball insights had transformed the game.

How can a team with a small budget maximize its chances of success? By finding players who can help them but who are undervalued. If every team knows how much a player could contribute, he'll be highly desired and cost more. If you have an insight others lack, you might sign that player at a discount.

In 2002, teams relied heavily on their scouts' opinions. Analytics offered another view. The Oakland A's worked from the premise, based in part on James's analysis, that statistics such as stolen bases, runs batted in, and batting average were flawed. Other statistics, like on-base percentage, could be better indicators of offensive success than speed and making con-

tact with the ball. In baseball, whether you get a hit or get walked, you are on base, with the potential to score a run. Hits and walks both contribute to on-base percentage but not batting average. Acting on this outlook enabled the A's to be competitive despite their small payroll. Today? Major League Baseball teams use baseball analytics, known as sabermetrics. Bill James is often considered a father of the field.

We'll look at just one of James's formulas, the "Pythagorean expectation," which proposes a connection between runs scored, runs allowed, and winning percentage. We'll test the formula on the 2019 Washington Nationals. First, we need the number of runs the team scored and allowed. In chapter 1, we drew on the trove of stats on baseball-reference.com. WolframAlpha.com, which we used in chapter 10, is even simpler; type in "2019 Washington Nationals," and it returns stats including the number of runs scored by the Nationals and by their opponents. The 2019 Nationals scored 873 runs and allowed 724.

James's Pythagorean expectation looks a bit like the Pythagorean theorem, which states $c^2 = a^2 + b^2$, where c is the length of the hypotenuse of a right triangle, and a and b the lengths of the two other sides. Be forewarned, the connection between James's formula and its classical namesake is quite loose. The Pythagorean expectation is similarly simple:

$$P = RS^2/(RS^2 + RA^2),$$

where RS is runs scored, RA is runs allowed, and P is the percent of games the team is expected to have won. For the 2019 Nationals, we get

$$0.5925 = 873^2/(873^2 + 724^2).$$

There are 162 games in the regular season, and 0.5925 × 162 is about 96. That's a pretty good estimate, as the Nationals won 93 games in the 2019 regular season.

....

Worksheet: Compute Like Bill James

Test James's Pythagorean expectation for another team. Pick a year and a Major League Baseball team.

1 Find runs scored by your team (RS): __
2 Find runs allowed by your team (RA): __
3 Compute $P = RS^2/(RS^2 + RA^2)$: __
4 Compute $P \times 162$: __

The result in line 4 is an estimate of how many games your team won in the season you chose. How close is it to the actual statistic? (For a year before 1962 or the short 2020 season, substitute the appropriate number of games.)

....

If we know how many runs a team scored and allowed in a given season, we most likely also already know how many games it won and lost and its winning percentage. So why use an estimate?

Because, by connecting runs and winning percentage, James's formula also lets us get a sense of individual players' contributions to a team.

Let's turn to the 2007 New York Yankees. They outscored their opponents 968–777, and their record, 94–68, was good enough to make the playoffs.

That season, third baseman Alex Rodriguez, also known as A-Rod, led the league with 156 RBIs. To see the impact of A-Rod's unforgettable year, suppose he had instead brought in only 103 RBIs, as he did the next season. (In both seasons, he was an All-Star and a Silver Slugger Award winner.)

The Pythagorean expectation estimates that the Yankees won $968^2/(968^2 + 777^2) = 0.608$ of their games in 2007. If A-Rod had brought in 53 fewer runs (156 − 103), the estimate becomes $(915^2)/(915^2 + 777^2) = 0.581$. The difference, 0.027, may seem small, but when we multiply it by the 162 games in an MLB season, we see a difference of about four games. In pro sports, every win is important!

Ed Porray

Position: Pitcher
Bats: Right · Throws: Right
5-11, 170lb (180cm, 77kg)
Born: December 5, 1888 in Atlantic Ocean
Died: July 13, 1954 (Aged 65-220d) in Lackawaxen, PA
Buried: Odd Fellows Cemetery, Lackawaxen, PA

Bill James's analytics helped make the 2002 A's competitive. If you can get four additional wins from one player and even one from another, it adds up and you increase your chances of going to the playoffs. Even better, the attributes the A's were looking for in players were undervalued in 2002. So the team was able to sign them for much less than the value they'd bring to the team and its record.

In the years since, other teams have integrated analytics into their approach to the game. And Bill James? He went on to serve as senior advisor on baseball operations for the Boston Red Sox. Quite a change from the night shifts at the cannery.

Analytic Toolkit

For player and team stats, we can visit sportsreference.com, as in chapter 6. Its baseball-reference.com database is an unparalleled archive of historical baseball stats, offering myriad directions for exploration. Consider the unremarkable career of Edmund Porray, who pitched 10⅓ career innings across three games in 1914, allowing eighteen hits and nine runs while not striking out anyone. Why note his page? Look past the game stats to his personal details. He was born at sea, and the site lists his birthplace as the Atlantic Ocean.

Personal Training—Workout #11

Bill James developed the Pythagorean expectation for baseball. Sports executive Daryl Morey adapted the formula to professional basketball. His formula to predict winning percentage is $P = PF^{13.91}/(PF^{13.91} + PA^{13.91})$, where PF denotes points for a given team and PA is points against. In  2019 the Toronto Raptors became the first Canadian team to win the NBA championship. Over the 2018–19 regular season, the Raptors scored 9,384 points while allowing 8,885 from their opponents.

Question: What would Daryl Morey's adaptation of the Pythagorean expectation for basketball predict for the Raptors' 2018–19 winning percentage? How many of the Raptors' 82 regular-season games would we expect them to have won?

Work on your calculation, then turn to the back of the book for a discussion of this question.

⓬

RACE USAIN BOLT

Jamaican sprinter Usain Bolt is, by many standards, the fastest human on earth. He holds world records in the 100 meter and 200 meter sprints and the 4 × 100 meter relay. He is an eight-time Olympic gold medalist—the only sprinter to win gold in the 100 meters and 200 meters at three consecutive Olympics (2008, 2012, and 2016). In the 2008 games, he set the 100 meter world record at 9.69 seconds.

In this chapter, you get to race Usain Bolt! You'll even have a chance of beating him! But with that hypothetical speed will come a warning about sports analytics.

While his 2008 Olympic run established a new world record, Bolt appeared on track to run an even faster race—until twenty meters before he crossed the finish line, when he looked at the other lanes and saw the strength of his lead. Most athletes celebrate moments after the end of the race. Bolt started celebrating before he finished. While Bolt is amazingly fast, even he slows down if he's celebrating while running.

We simply don't know how fast Bolt might have run the 100 meters that day. His coach suggested his time could have been 9.52 seconds or better, and physicists have confirmed that's plausible.

To race Usain Bolt and his 9.69 second record we won't go to a track. As with our other simulations, we'll compete right where we are. So, what assumptions will we need to make this work?

Let's start out with three:

1 The number of steps you can take in a given time is the same whether you are running in place or on a track.

2   Your stride length will be 1.35 times your height.

3   Stride length never changes.

Want a more conservative estimate? The 1.35 multiplier does come from research published in the early 1970s—but other researchers calculated stride length to be only 1.14 to 1.17 times the sprinters's height. I picked the higher value to give you a chance at winning! If you want a harder race, replace 1.35 with 1.14 in the computations.

Worksheet: Racing Usain Bolt

Before chasing Bolt's record, you need to answer two related questions:

- How long is your stride?
- How many strides, assuming constant stride length, will you need to complete 100 meters?

The computations are simple:

- stride length = (height in meters) × 1.35
- number of strides = 100/(stride length in meters)
- round the number of strides up to the nearest integer

I'll use myself as an example: I'm 6 feet tall, or 1.83 meters. So my stride would be 1.83 × 1.35 = 2.47 meters. And how many 2.47 meter strides will it take me to cover 100 meters. Well, 100/2.47 = 40.4, which I round up to 41 strides.

Once you know the number of strides, the race is ready. Running in place, complete that many steps in under 9.69 seconds and you beat Usain Bolt! Run fast and, of course, be careful of celebrating early!

Did you beat Usain Bolt or almost win? If you and a few friends try, some will likely do so. But wait! How are we competing with, let alone beating, Bolt?

Let's look again at our assumptions. We assumed a consistent stride length. This clearly does not hold true for an entire physical race. An athlete simply cannot take the same strides coming out of the starting blocks as later in the race. Our model set my stride length at 2.47 meters (or 2.09 meters if I use the more conserva-

tive 1.14 multiplier) and assumes that I could take such strides at the same rate I can run in place. A little thought debunks the assumption. I obviously can take forty-one "strides" in place much faster than forty-one maximally extended strides down a track. These skewed assumptions partially explain how I can "beat" Bolt.

This is an important analytics lesson. Surprising results can yield groundbreaking insight. But often they mean you need to check your work and assumptions. My beating Bolt would be unforgettable, but it is also simply impossible. It can be fun to set parameters that make the race appear competitive. But such a model is wildly distorted.

Herein lies a moral: Be wary of your assumptions.

Throughout this book, we've made simplifying assumptions, which have facilitated rough estimates of probabilities. Steph Curry was presumed to be about 43% likely to hit any three-point shot. Wouldn't that depend in part on the defender? Alex Morgan's goals/cap rate was presumed to hold for every game. Wouldn't that, too, depend on the strength of the opposing team's defense?

Want a better estimate? Complicate your assumptions! Remove some of the distortion inherent in simplication. Bear in mind, though, that less simplified models may require more advanced mathematics or statistics.

If you don't need the accuracy, then a simplified model may serve your purpose. But be mindful that truly surprising results, like beating Usain Bolt, may mean the oversimplified has lapsed into the unrealistic.

Personal Training—Workout #12

Let's take a run at another world record, set in 1988 at the Seoul Olympics by Florence Griffith Joyner. Rework the model in this chapter, and see what it would take to beat Flo-Jo's record in the 200 meters—21.34 seconds.

Question: Assuming the same 1.35 ratio of stride length to height as we did in our race with Usain Bolt, how many absolutely consistent strides would someone 5 feet, 9 inches tall need to make in under 21.34 seconds to beat Flo-Jo's world record–setting 200 meter sprint?

Calculate your answer, then turn to the back of the book for a discussion of this question.

13

GET IN THE GAME

When you witness the unlikely, the tools in this book can give you a better sense of whether it's historic, allowing you to get more fully into the game. There is another way you can get in the game—becoming part of the team by contributing from the sidelines. How? Let's start with my story.

In the fall of 2013, three students came to my office and told me they wanted to create analytics for the Davidson College men's basketball team. I'm a college professor—just imagine how thrilled I was as a teacher. Standing in front of me were students looking to do extra mathematics on their own time!

The men's basketball program was already highly successful. Steph Curry had played for Davidson in his college years. The students understood that our efforts might not generate useful insights, but they were eager to begin. Our first step was an important one. We contacted the coaching staff. The coaches were interested, and over the next several months we met weekly to prepare for the season.

Week by week, our team posited ideas, prototyped analytics, and shared what we developed. And, crucially, we got feedback. We learned to understand the team's needs, and the coaches learned what sorts of tools we could create and questions we could answer.

Not everything we developed worked. Sometimes, the coaches told us, a conclusion simply wasn't "coachable." For instance, knowing the team is ranked eleventh in the country in a type of defense is interesting. But knowing how the team can maintain that ranking, or improve it based on what higher-ranked teams are doing differently, can actually impact coaching.

When a coach said something wasn't coachable, we learned that the next question should always be "Why?" Sometimes an uncoachable idea simply needed to be reframed to be very useful. Even ideas that were scrapped were far from useless, as they enabled us to better understand the coaches' needs.

By tip-off of the first game, we had our tools ready. Soon, we were rolling through the season. The coaches depended on us. Our analytics became important in their game preparation. Our work had four components: We charted shots, recording position, shooter, and point value (later adding such features as time on the shot clock and shot type). We visualized the shot charts by point value (where do we miss?) and by athlete (where does one player make or miss?) over any combination of games. We prepared player and team scouting reports. And we logged statistics for every lineup

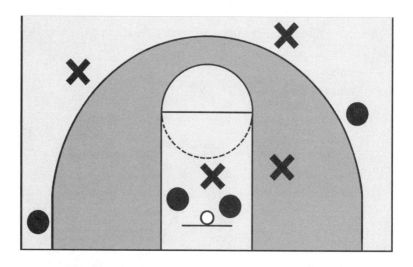

that took the court. This allowed us to analyze not only every five-player combination but also smaller combinations, such as pairings of a particular guard and forward.

The program was a success. We supplied analytics and, as the coaches often noted, became part of the team, playing from the sidelines. In 2014 we had four students. Jump to now, and the sports analytics group numbers a hundred. We supply analytics to coaches in a range of sports: men's and women's basketball, baseball, men's and women's soccer, volleyball, football, lacrosse, and swimming. When we meet with a new coach, our first question is "What do you do with numbers and what do you wish you could do?" Then, we listen and learn. Soon afterward, the group meets, with a lot of ideas and analytics being written.

As we develop our analytics, here are our guiding principles:

- **Coachable.** Share insights that coaches can act on.
- **Consumable.** Present insights in formats that can be quickly synthesized.
- **Understandable.** Insights are an informed opinion. Be sure a coach can understand what led to the work.

Since those early days, the scope of our work has expanded well beyond our college. We field questions from national media, from ESPN to the *New York Times*. We aid professional teams in the NBA, NFL, and NASCAR. We've also helped the US Olympic and Paralympic Committee. In each setting, we focus on actionable, consumable, and understandable analytics. From college coaches to work seen on television, our goal is to be unforgettably dependable.

Now, how do *you* become part of a team? First, watch sports and pose questions. Yes, from your couch you can train to become part of a sports team. As we've done throughout this book, derive estimates to answer your questions. Then think of new ones! Looking for a first step? Write down whatever questions came to mind as you read this book. Or search the internet for articles and books on what others have done. Existing work can be a springboard for your own ideas. And then . . .

Jump in. Want to work in sports analytics? Think of a question related to one you already know how to solve. Then, just start. If you wait until you're sure you are ready, you may never make that first move.

Step into the unknown. When ESPN would call and ask for help, I rarely knew right off how to approach the problem. Part of analytics is learning to stand in the unknown waiting for that next step, which may not come instantly.

Keep learning. You may pose a question that requires math, stats, or computing that are beyond what you know. There are numerous resources online. Continue to study and add to your skills. Be willing to ask someone who might know more for help!

So take the leap! Creating and answering your own questions is practice in sports analytics. As the great gymnast Simone Biles has said, "Practice creates confidence. Confidence empowers you." Listen to your thoughts, as they can create direction, insight, and confidence. As Biles also stated, "If you're having fun, that's when the best memories are built." So remember, too, to have fun.

Keep your tools handy—a die, a coin, a calculator, a computer. Take note when you hear "Wow!" during a sporting event. Analyzing surprising moments can make them all the more *unforgettable*. And

it's *improbable* that all those computations won't get you more fully into the game!

Personal Training—Workout #13

Want to jump into sports analytics? You'll need data. While professional sports are extensively documented in public and private databases, many other levels of sport simply don't have as much data. Recording data can be part of your analytics. The Cats Stats group began by charting shots for basketball. The half-court diagram comes from an app we developed for high schoolers to record their data. A roster is loaded. Then, the location of each shot is recorded, along with the name of the shooter, type of shot, and point value. What else would you record? The data you record opens doors to the type of analysis you can explore.

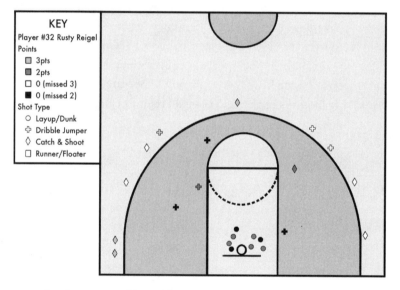

Write down your ideas, then turn to the back of the book for a discussion of this question.

ACKNOWLEDGMENTS

First and foremost, I thank Tanya Chartier. Many of our walks have involved my outlining and reoutlining talks and eventually this book. Thank you, Tanya, for the many ways you encouraged my vision. I'm grateful to Ansley Earle for her illustrations but even more for her insight on how to approach the book. Joseph Calamia, my editor at University of Chicago Press, thank you for your creativity, engagement, encouragement, and keen insight. The friendship that grew via this book is a highlight of the project. I thank Roland Minton for his careful review of the manuscript. His input played an especially important role in shaping chapter 10. Roland, thank you for the many ways you've enhanced my work through the years. Thanks to Josh Tabor for his answers to my many questions and for coauthoring the text *Statistical Reasoning in Sports*, which inspired portions of this book. Thanks to Tyler Heaps, of the US Soccer Federation, for his input and help. I want to thank Amy Langville, with whom I first launched into sports analytics. Amy, thank you your friendship and collegiality. Thanks to Rahul Prakash and Drew Dibble for their reading of the book; I appreciate your input and affirmation. I also thank Jan and Myron Chartier for their support when this book was simply a vague image in my mind. Thanks to Melody Chartier who affirmed each stage of the book. Thanks for MaryJo Johnson for her encouragement each time we discussed this book. Thanks to Noah and Mikayla Chartier, who were often my sounding board for new ideas—when they laughed or their eyes lit up, I ran upstairs to begin writing! Thanks to the Davidson College men's basketball coaches who took a first step alongside me into the analytics unknown; a highlight of my professional career is being "part of the team." Finally, I want to thank the many members of Cats Stats, my sports analytics group at Davidson College. You enrich my life as a professor, you push my work into new territories, and you walk with me into the unknown when faced with a new question or problem from a coach, professional sports team, or national media outlet. Charts like those on pages 4–6 were prepared using roughViz software by Jared Wilber (https://github.com/jwilber/roughViz). It is difficult, if not impossible, to thank everyone who played a part in this book, from colleagues to students to friends. For those mentioned and unmentioned, may you see your encouragement and the results of our conversations in these pages.

FURTHER READING

Chapter 1: Unforgettably Unbelievable

Unforgettable moments are often improbable moments. In this chapter, we looked at players who have outlier games, seasons, or careers. Once you've gathered data (I mentioned two baseball databases: baseball-reference.com, which is part of www.sports-reference.com, and SeanLahman.com), this short tutorial, on topics from sorting in a spreadsheet to using statistical z-scores, can help you identify outliers:

Jim Frost. "5 Ways to Find Outliers in Your Data." Statistics by Jim. Accessed July 13, 2021. https://statisticsbyjim.com/basics/outliers/

I noted that graphs can be misleading. These authors explain some of the ways graphics deceive:

Becca Cudmore. "Five Ways to Lie with Charts." *Nautilus*. November 6, 2014. Accessed July 13, 2021. https://nautil.us/issue/19/illusions/five-ways-to-lie-with-charts

Alberto Cairo. *How Charts Lie: Getting Smarter about Visual Information*. New York: W. W. Norton, 2020.

Christopher Ingraham. "You've Been Reading Charts Wrong: Here's How a Pro Does It." *Washington Post*. October 14, 2019. Accessed July 13, 2021. https://www.washingtonpost.com/business/2019/10/14/youve-been-reading-charts-wrong-heres-how-pro-does-it/

Chapter 2: Shoot 3's like Steph Curry

Stephen Curry's shooting accuracy is historic. The performance discussed in this chapter has led many to say he had a "hot hand" that evening. We showed that hitting 11 of 13 three-point attempts was highly unlikely in the context of being able to say on a given night, "Stephen Curry will shoot 11 of 13 or better tonight." Was Curry hot? Is there a hot hand?

Many, including Nobel Prize winner Daniel Kahneman, have claimed the hot hand is an illusion. Yet, research has uncovered a fundamental flaw in the work that launched the anti–hot hand belief.

Joshua Miller and Adam Sanjurjo. "Momentum Isn't Magic: Vindicating the Hot Hand with the Mathematics of Streaks." *Scientific American*. March 28, 2018. https://www.scientificamerican.com/article/momentum-isnt-magic-vindicating-the-hot-hand-with-the-mathematics-of-streaks/

Ben Cohen. *Hot Hand: The Mystery and Science of Streaks*. New York: William Morrow, 2021.

Chapter 3: Dicey Hitting Streak

Want to read more about the DiMaggio brothers' rivalry?

Scott Ostler. "Sibling Rivalry? DiMaggios Had No Peers." *SFGATE*. July 15, 2015. Accessed July 13, 2021. https://www.sfgate.com/sports/ostler/article/Sibling-rivalry-DiMaggios-had-nopeers-6387146.php

Here's an argument for intentionally walking a big hitter like Joe DiMaggio:

Steven Goldleaf. "The Case for the Intentional Walk: Articles." Bill James Online. March 20, 2021. Accessed July 13, 2021. https://www.billjamesonline.com/the_case_for_the_intentional_walk/

In the 1940s pitchers were often booed, even by home fans, for not throwing strikes, and the significance of walks—and consequently Ted Williams's 84-game on-base streak—was long undervalued. This article offers a more recent assessment:

Ken Schultz. "Ted Williams' Underappreciated On-Base Streak." Boston. March 02, 2017. Accessed July 13, 2021. http://boston.locals .baseballprospectus.com/2017/03/02/ted-williams-underappreciated-on -base-streak

Chapter 4: Racking Up the Wins

Much has been written about the remarkable streaks discussed in this chapter. Here are two articles to start with:

Matt Bonesteel. "How Big Was Mississippi State's Upset of U-Conn? In Vegas Terms, Huge." *Washington Post*. April 1, 2017. Accessed July 13, 2021. https://www.washingtonpost.com/news/early-lead/wp/2017/04/01/how -big-was-mississippi-states-upset-of-u-conn-in-vegas-terms-huge/
Doug Ammon. "Delle Donne's Unprecedented Free-Throw Shooting Continues to Amaze." WNBA.com. July 27, 2017. Accessed July 13, 2021. https:// www.wnba.com/news/elena-delle-donne-free-throw-foul-shooting/

By the end of the chapter, we knew the formulas for permutations and combinations. This blog and video offers more practice:

Brett Berry. "Combinations vs Permutations." Medium. June 14, 2017. Accessed July 13, 2021. https://medium.com/i-math/combinations -permutations-fa7ac680f0ac

Chapter 5: Unbreakable Tennis

Our analysis of Pete Sampras and Andre Agassi relied on their career stats. To find stats on a player like Novak Djokovic visit the player's page on ultimatetennisstatistics.com:

"Ultimate Tennis Statistics—Novak Djokovic." ATP Tour. Accessed July 13, 2021. https://www.ultimatetennisstatistics.com/playerProfile ?playerId=4920

In this chapter, we discussed programming simulations. We could use languages such as Python or R or spreadsheets like Excel or Google Sheets. These articles consider two examples:

Lloyd Danzig. "An Intro to Monte Carlo Simulation for Sports Betting Risk Management (in Excel)." Medium. July 19, 2020. Accessed July 13, 2021. https://medium.com/@lloyddanzig/an-intro-to-monte-carlo-simulation -for-sports-betting-risk-management-in-excel-c951a144f13a

Michael Emmert. "Texas Hold'Em Simulator." Medium. June 15, 2020. Accessed July 13, 2021. https://towardsdatascience.com/texas-holdem -simulator-e2b83537b133

Chapter 6: Improbable NFL Playoff

This chapter dealt with an improbable score at the end of regulation play in an NFL playoff game. This article looks at other improbabilities in football (and note its use of "Pythagorean games," a variation of the Pythagorean expectation discussed in chapter 11):

Bill Barnwell. "Ranking the Most Unlikely Super Bowl Teams Ever, and Where the 2019 49ers Land." ESPN. January 23, 2020. Accessed July 13, 2021. https://www.espn.com/nfl/story/_/id/28538404/ranking-most -unlikely-super-bowl-teams-ever-where-2019-49ers-land

One tab on the www.sports-reference.com site links to stathead.com, which covers football, baseball, hockey and basketball. Stathead returns some results according to your criteria for free and many more with a subscription.

Chapter 7: FIFA Octopus Oracle

For more on the more or less improbable predictions of Paul the Octopus and Barack Obama:

Emily Shire. "The Amazing Tale of Paul the Psychic Octopus: Germany's World Cup Soothsayer." *Daily Beast*. July 12, 2014. Accessed July 13, 2021. https://www.thedailybeast.com/the-amazing-tale-of-paul-the-psychic -octopus-germanys-world-cup-soothsayer

Felippe Rodrigues. "Reviewing Every Barack Obama March Madness Bracket." Game Plan. March 20, 2019. Accessed July 13, 2021. https://web .northeastern.edu/gameplan/2019/03/20/reviewing-every-barack-obama -march-madness-bracket/

For coaches seeking insight on the use of *p*-values:

Craig Pickering. "Understanding P-Values: A Practical Guide for Coaches." SimpliFaster. Accessed July 13, 2021. https://simplifaster.com/articles/understanding-p-values-practical-guide-coaches/

Chapter 8: Super-sized Super Bowl TD

Was the attention paid to William Perry's Super Bowl touchdown merely a by-product of his team's broader notoriety? Here's one take on his team's place in NFL history:

Matt Reagan. "1985 Chicago Bears: The Greatest Team Ever." Bleacher Report. March 10, 2009. Accessed July 13, 2021. https://bleacherreport.com/articles/136752-1985-chicago-bears-the-greatest-team-ever

Here's a bit more reading on the normal distribution:

"Normal Distribution." Math Is Fun. Accessed July 13, 2021. https://www.mathsisfun.com/data/standard-normal-distribution.html

Other distributions are also relevant. Here's a look at applications of the power law:

Brian Burke. "Earthquakes, Kevin Bacon, the Financial Crisis, and Pro Bowl Selections." Advanced Football Analytics. Accessed July 13, 2021. http://archive.advancedfootballanalytics.com/2009/04/earthquakes-kevin-bacon-financial.html

Chapter 9: Scoring Confidence

For context and analysis of Alex Morgan's hundredth goal:

"100: Alex Morgan Joins Exclusive WNT Club." US Soccer. April 5, 2019. Accessed July 13, 2021. https://www.ussoccer.com/stories/2019/04/100-alex-morgan-joins-exclusive-wnt-club

For more detail on confidence intervals:

Saul Mcleod. "What Are Confidence Intervals?" Simply Psychology. June 10, 2019. Accessed July 13, 2021. https://www.simplypsychology.org/confidence-interval.html

Chapter 10: Tiger's Consistency

For more on consistency in golf, and the wider world of golf analytics:

Roland Minton. "ShotLink and Consistency in Golf." Access July 13, 2021. https://apps.roanoke.edu/minton/consistency.pdf
Roland B. Minton. *Golf by the Numbers: How Stats, Math, and Physics Affect Your Game.* Baltimore, MD: Johns Hopkins University Press, 2012.

Chapter 11: Moneyball Analytics

For more on baseball analytics and its effect on baseball:

Benjamin Baumer and Andrew. Zimbalist. *The Sabermetric Revolution: Assessing the Growth of Analytics in Baseball.* Philadelphia: University of Pennsylvania Press, 2013.

For a popular account of how Bill James's ideas entered Major League Baseball, later made into a movie:

Michael Lewis. *Moneyball: The Art of Winning an Unfair Game.* New York: W. W. Norton, 2013.

Or go straight to the source:

"Welcome to Bill James Online." Bill James Online. Accessed July 13, 2021. https://www.billjamesonline.com/

For the state of analytics, at least in 2015, within the NBA, MLB, NHL, and NFL:

Ben Baumer, Kevin Pelton, Craig Custance, and Kevin Seifert. "The Great Analytics Rankings." ESPN. February 23, 2015. Accessed July 13, 2021. https://www.espn.com/espn/feature/story/_/id/12331388/the-great-analytics-rankings

Chapter 12: Race Usain Bolt

How fast can humans run? Our understanding of this question comes, in part, from considering the case of Usain Bolt:

"How Does Usain Bolt Run So Fast?" Runners Home. Accessed July 13, 2021. https://www.run3d.co.uk/news/how-does-usain-bolt-run-so-fast

Teal Burrell. "How Fast Can Humans Go?" *Discover*. July 2, 2018. Accessed July 13, 2021. https://www.discovermagazine.com/health/human-speed

Chapter 13: Get in the Game

For more on the Cats Stats sports analytics group at Davidson College:

Seth Berkman. "Davidson Math Students Lend a Hand to Basketball Team." *New York Times*. March 14, 2015. Accessed July 13, 2021. https://www.nytimes.com/2015/03/15/sports/ncaabasketball/davidson-math-students-lend-a-hand-to-basketball-team.html

Martin Kessler. "Math Students Give Davidson Athletics an Edge." *Only a Game*. NPR. January 17, 2020. Accessed July 13, 2021. https://www.wbur.org/onlyagame/2020/01/17/davidson-college-cats-stats-advanced-analytics

With these resources and others you will find as your own interests and questions arise, you are ready to dive into sports analytics, get in the game, and uncover the unforgettably improbable moments in the ongoing history of sports!

PERSONAL TRAINING —ANSWERS

The personal training workout at the end of each chapter poses a question. Try to answer it before reading the discussions below.

Workout #1: The first step in this workout is finding data on successive world records in the women's 1500 meters freestyle. One quick resource is Wikipedia: https://en.wikipedia.org/wiki/World_record _progression_1500_metres_freestyle. For a more formal analysis, you should look for other sites to verify the numbers. Looking at world records from 1922 until today, we see that Janet Evans's record, set in 1988, stood for 7,022 days, considerably longer than any of the others. How much longer? There are a variety of ways to answer this question. We can note that the next longest-held

record was set by Ragnhild Hveger on August 20, 1941, and stood for 5,085 days. The third longest was set by Kim Linehan in 1979 and held for 2,903 days before being broken by Evans in 1987. We could graph the data, or we might decide that looking at the numbers is sufficient to spot outliers. Spreadsheets

quickly tally the number of days between two days. You can also use WolframAlpha (type in the query "number of days between 6/17/2007 and 3/26/1988") to compute the length of Evan's record as 7,022 days.

Workout #2: During the 2020–21 NBA season, Stephen Curry's field goal shooting percentage was 48.2%—close enough to 50% that a coin flip provides a reasonably good model. A quick approach to

estimating the improbability of Curry's going 2 for 16 in a game is to flip a coin sixteen times and see if you only flip two heads. Or you can break the analysis down by types of shot. Curry made 1 of 6 two-point attempts and 1 of 10 three-point attempts. His two-point shooting percentage for the regular season was 56.9% and his three-point shooting percentage was 42.1%. To approximate these chances with flips of a coin, we need simple fractions close to these values. With four flips, there are sixteen possible outcomes ($2 \times 2 \times 2 \times 2$, or 2^4); $9/16 = 0.5625$, which is close to Curry's two-point shooting percentage, and $7/16 = 0.4375$, a good approximation of his three-point percentage. So you can flip a coin four times and, for a two-point attempt, make nine of the possible outcomes count as a made shot; for a three-point attempt, seven of the possible outcomes represent a hit.

Workout #3: The Houston Rockets had won about half their games before late January 2008. If we assume their chance of winning any subsequent game was 0.5, then the probability of winning twenty-two straight games was 0.5^{22}—about 0.00000024, or 1 in 4.2 million. What if we assume that the team's final record, 55–27, better reflects its ability? That makes the chance of the Rockets winning any game 0.67, and the probability of winning twenty-two straight games 0.67^{22}, or about 0.00015. This is about 625 times more likely—but still only about 1 in 6,700. Note that we are assuming independence—that winning any one game does not influence the outcome of the next game. This isn't necessar-

ily the case. For example, winning streaks can lead to larger crowds, which may lead to more wins. Even so, assuming independence is a helpful place to begin and can help with basic estimates, which can springboard into more sophisticated models.

Workout #4: For the 2019–20 NHL playoffs, two "hub" cities were selected from the seven Canadian cities with NHL arenas. How many ways could two cities have been selected? We have a group of seven items (cities) from which we are choosing two, and order

doesn't matter (since selecting Edmonton and Toronto is the same as selecting Toronto and Edmonton). So our answer is C(7,2) = 7 × 6/(2!) = 21.

Workout #5: In her 2012 Wimbledon semifinal against Victoria Azarenka, more than a third of Serena Williams's serves were aces. What's the probability of a player with an ace percentage of 14.4% matching her achievement of 24 aces in 71 serves? One approach is to treat this as a Bernoulli trial: $C(n,r) = n!/((n - r)! \times r!)$. With 0.144 as the probability of success for each serve, the overall probability is $C(71,24) \times (0.144)^{24} \times (1 - 0.144)^{(71 - 24)} = 0.000022$, or about 1 in 45,000.

Another approach is to write a computer simulation. Here is how a computer can serve like Serena Williams:

- Set success counter to 0. Then repeat the following steps a million times.

- ○ Set the ace counter to 0. Then repeat the following steps 71 times.
  - Pick a uniform random number between 0 and 1.
  - If the number is less than 0.144, the serve was an ace. Add one to the ace counter.
  - ○ Did you serve 24 aces? If so, add one to the success counter.
- Divide your total number of successes by one million and you have your estimate for the probability we just found treating this as a Bernoulli trial.

Why run the simulation a million times? The underlying probability is so small that ten thousand repetitions produce an estimate of a zero probability on many runs. In fact, we'll do best to make multiple runs of a million tries. When you see approximately the same value on multiple runs, you can trust your estimated value via simulation.

Workout #6: During the 2016 MLB season, there were 165,561 at-bats and 5,610 home runs, making the probability of hitting a

home run 5,610/165,561 = 0.0339. We want to determine the best approximation of that probability in a simulation where the roll of three dice determines the outcome of an at-bat. There are 216 possible outcomes of rolling three dice (6 × 6 × 6, or 6^3). Multiplying 0.0339 by 216 yields 7.32, and indeed, 7/216 = 0.0324 is the closest match. You can check 6/216 = 0.0278 and 8/216 = 0.0370 to verify that 7/216 is our best option.

Did your answer count only seven outcomes of rolling three dice as home runs? If so, you found an optimal combination of rolls to approximate the probability of hitting a MLB home run in 2016. There are a variety of ways to come up with seven options when rolling one die three times. For example, we could call it a home run if we get 6's on both the first and second roll (plus any of six possibilities on the third) or if we roll three 5's. This gives us a homer for rolling 6-6-1, 6-6-2, 6-6-3, 6-6-4, 6-6-5, 6-6-6, or 5-5-5. Grab a die, and play ball!

Workout #7: In 2016, President Barack Obama correctly predicted 23 of 32 games in the first round of the 2016 men's March Madness tournament. To determine the probability that random chance (a coin flip) could have done as well or better, you can use a Bernoulli trial assuming a probability of success of 50%. Either use the formula from chapter 4, or take a shortcut by going to WolframAlpha.com and typing in "probability of at least 23 heads out of 32 flips," which returns a probability of about 0.01. (With random picks, there is an equal chance of picking at least 23 of 32 games wrong.)

Workout #8: Kareem Abdul-Jabbar's height is recorded as 7 feet, 2 inches. If we plot NBA players' heights over sixty-seven seasons beginning in 1950, we find a fairly normal curve. The mean

height is 6 feet, 6 inches, with a standard deviation of 3.65 inches. It will be easier to work with
heights in consistent units and decimal form.
In feet, Abdul-Jabbar's height is 7.167, the mean
NBA height is 6.5, and the standard deviation is
0.3042. So his height is $(7.167 - 6.5)/0.3042 =$
2.19 standard deviations above the average. Al-
though not part of the question, this converts to
an estimate that just over 98.5% of NBA players
have been shorter than Kareem Abdul-Jabbar. So
even though he wasn't the tallest, he's very tall,
even in the NBA.

Workout #9: Over her international soccer career,
Abby Wambach scored 184 goals in 256 caps, giving her a goals/cap rate of 0.719. To simulate her
scoring, a computer could follow these steps:

- For each game, pick a uniform random number between 0
  and 1.
- If the number is less than 0.719, Abby Wambach scores in the
  game. Else, she doesn't.

- Play 256 games, tally the number of games in which she scores, and compute the goals/cap rate.
- Repeat ten thousand times.

When the computer plays 256 games ten thousand times, we find that 95% of the simulations have Wambach's goals/cap rate between 0.66 and 0.78.

Workout #10: In chapter 10's worksheet, a PGA golfer made the cut when a roll of four dice summed to at most 14. We determined that 721 of the 1,296 possible outcomes of a four-dice roll summed to 14 or less, and so calculated a $(721/1296)^{30}$ chance—about 1 in 43 million—of making thirty straight tournament cuts. In our final model for Tiger Woods, he made the cut when the four dice summed to at most 18. There are 1,170 ways of rolling a sum of 18 or less with four dice, which improves his chances of making the cut in thirty straight tournaments to $(1170/1296)^{30}$, or about 1 in 21. Skill and consistency massively improve the odds of a streak!

Workout #11: Daryl Morey adapted Bill James's Pythagorean expectation to predict winning percentage in professional basketball as $P = PF^{13.91}/(PF^{13.91} + PA^{13.91})$. Plugging in stats numbers for the Toronto Raptor's 2018–19 regular season—9,384 points for (PF), 8,885

points against (PA)—we find the Raptors could be expected to have won 0.6814 (68.14%) of their games. There are 82 games in the NBA regular season, making the expected number of wins $82 \times 0.6814 = 55.9$. In fact, they won 58. John Hollinger, another prominent figure in basketball analytics, tweaked the Pythagorean expectation for basketball using a different exponent, 16.5—which is a better predictor in the case of the 2018–19 Raptors: $9,384^{16.5}/(9,384^{16.5} + 8,885^{16.5}) = 0.7113$; $82 \times 0.7113 = 58.3$.

Workout #12: In this workout, we apply the model that told us how many strides we would have to make in less than 9.69 seconds to beat Usain Bolt's record in the 100 meters to a longer race. Here,

our competitor is 5 feet, 9 inches tall and looking to beat Florence Griffith Joyner's time of 21.34 seconds in the 200 meter sprint. First, we convert 5 feet, 9 inches (5.75 feet) to 1.75 meters, then multiply by 1.35 for a stride length of 2.37 meters. To cover 200 meters taking strides with a constant length of 2.37 meters would require $200/2.37 = 84.4$, which we round up to 85 strides. If our 5 foot, 9 inch runner can complete those 85 strides in under 21.34 seconds, then she beats Flo-Jo, thanks again to an overly simplified model!

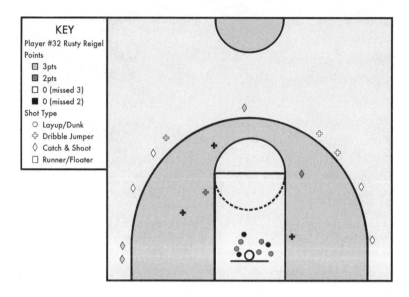

Workout #13: The Cats Stats sports analytics group at Davidson College began by charting shots for basketball. We now also chart data for football, baseball, lacrosse, soccer, and other sports. For basketball, we record the information noted at the end of the chapter—shot location, name of shooter, type of shot, and point value—plus shot clock, assist, and game clock information, as well as other data. When our student analysts record live at a basketball game, it now takes three of them to enter all the data. In the beginning, it took only one student. So think about what you want to record. Once you've charted a handful of games, you can begin looking for patterns. In time, you can look for patterns based on individual players, groups of teammates, opponents, time in the game, or location on the court. What else comes to mind? Listen to your ideas, and you may develop a new direction offering new insight. You may develop the unforgettably new sports analytic.